ENVIRONMENTAL STUDY IN HIMALAYAN REGION

ENVIRONMENTAL STUDY IN HIMALAYAN REGION

By

D. S. Jalal • B. R. Pant
R. C. Joshi

ANMOL PUBLICATIONS PVT. LTD.
NEW DELHI - 110 002 (INDIA)

ANMOL PUBLICATIONS PVT. LTD.
4374/4B, Ansari Road, Daryaganj
New Delhi - 110 002

Environmental Study in Himalayan Region

First Edition, 2001

ISBN 81-261-0936-X

PRINTED IN INDIA

Published by J.L. Kumar for Anmol Publications Pvt. Ltd., New Delhi - 110 002 and Printed at Mehra Offset Press, Delhi.

Contents

Preface

Complex and intricate problems arising from land degradation in Himalaya have made environmentalists conscious of the immediate need of conservation and optimal use of the land resources. On account of the formidable problems issuing from the stagnant and deteriorating agricultural conditions, the ever increasing pressure of population on limited land resources, the poor condition of livestock wealth, the serious consequences of thoughtless deforestation, indiscriminate grazing and consequent soil leaching, erosion, landslides, rock-falls, floods and siltation, have made the need of landuse planning more than ever necessary.

Himalaya is a geographical-cultural continuum having innumerable regional variations. Each region boasts of its own individuality, asserts its uncommon characteristics, socio-economic and environmental problems, hence requiring a special treatment in its own way. A comprehensive survey of these problems should ultimately be conducted individually besides taking the country as a whole. Plans for improvements should therefore be based on carefully prepared local surveys. Present work is an attempt at filling a part of this need.

An attempt is made here at presenting a vivid account of the general and agricultural landuse of Kotadun which is characterized by highly undulating topography to the north and represents the landuse characteristics of Outer Kumaun Himalaya. Landuse survey even in the verification of records, needs detailed field investigation for which the researchers visited to the study area. For further details three sample villages are selected at random, from three well identified physical strata of dunfloor or valleys, midslopes and uplands. The conclusion about agricultural efficiency, land evaluation, nutrition and deficiency diseases are based on the findings pertaining to those villages.

Authors' obligations are numerous. First of all we express our deep sense of gratitude to Dr. D.D. Pant, Professor Emeritus, Physics Department, Kumaun University, Nainital, Dr. K. S. Valdiya, former Professor and Head, Geology Department, Kumaun University, Nainital and Dr. J. S. Singh, Professor of Botany, Benaras Hindu University, Varanasi for their keen interest, inspiration and guidance at every step of the project work. We are also thankful to late Dr. S. L. Shah then Project Coordinator, Kosi Valley Action Research Project, Central Himalayan Environment Association, Nainital for extending his valuable suggestions. We record our indebtedness to the Vice-Chancellor, Kumaun University, Nainital and Campus Director for providing space and other basic facilities to establish the Optimal Landuse Research Laboratory in D. S. B. Campus of the University. Thanks are also due to Dr. L. S. Khanka, Dr. (Smt) Sudha Samant, Dr. N. S. Bora, Dr. A. Pande, Kunwar Singh and Vijay Singh for their kind help during the present work.

Grateful acknowledgement is conveyed to the Ministry of Environment, Forest and Wildlife, Department of Environment, New Delhi for generous sanction and funding of the project. Present book is an outcome of the Annual Progress Reports submitted to this funding department.

The area of the study is away from the University Headquarters and the maps and records maintained in various offices are confidential in nature and are far from availability. In this regard the investigators are thankful to the District Magistrate, Nainital, S.D.M., Kashipur and Peskars Ramnagar, Kaladhungi and Nainital for making them all the required information available. We are particularly thankful to the authorities concerned for allowing us 10 ha of public waste land in Patkot in order to execution of action oriented programme. We are also thankful to the owner cultivators of Patkot and Saunjala for allowing these programme in 3.6 ha and 1.2 ha of their respective private lands. At the last but not the least we records our sincere thanks to residents, social workers and officials of Kotadun who co-operated the investigators in all respect of research as well as action programme operations.

D. S. Jalal,
B. R. Pant
R. C. Joshi

List of Tables

List of Figures

Chapter 1
Introduction

1.1. THE PROBLEM

The ever increasing pressure of population, deteriorating environmental conditions and increasing value of land signify to study every use, misuse, overuse and underuse of land. Much can be said about the basic importance of optimal use of our limited land resources. According to Stamp (1964) food, shelter, work, recreation, movement and security are the six basic needs of man which involve the use of land for their fulfilment. There are always reasons whether of natural or cultural factors that have determined the use of land in a particular area. The present day landuse of any area is certainly an outcome of hundreds of years of ecological process and human endeavour.

The use of a particular piece of land varies according to the needs of society in changing space of time. Man with his knowledge and skill has converted the land into large agricultural farms, big cities, industrial complex etc. Of course, the nature of land has to play very important role before man's occupancy. There are areas where agriculture is not at all possible or is uneconomic in certain areas, some land is not fit for the development of large cities, industrial complexes or even for the purpose of activities like grazing, forestry, quarrying and constructions. So the landuse of any area is a complex phenomenon and is the outcome of the complete system of a region.

Undoubtedly, prosperity of any region, growth of its economy and improvement in the living condition of its people much depends upon the proper utilization of the land available. Since, the land is scarce and limited, its misuse can yield very drastic horrified results to the physical as well as the cultural environment, while a just, wise, scientific and proper use of every piece of land can facilitate to grow a region faster and ably manage its future plans. The limited area of land can in no way be expanded while the population depending on that particular land grows faster which spares no alternative to accommodate this extra burden but to utilize the land properly and optimally with scientific endeavours. The definition of proper or optimal use, however, varies from one piece of land to another.

Thus, it is of utmost importance to determine the present uses of land in a region and to ascertain, if possible how it has come into being. The future development and prosperity according to Stamp (1960), must start from the present existing pattern. The landuse planning, thus, seeks for the factors which are operative, to think in terms of potential use and what land ought to be used for.

The basic need today is to think about the reorganization of existing land resources and then to execute plans for its improvement, optimization and re-orientation. The optimum situation can only be achieved through scientific landuse planning. The landuse maps play an important role in this regard in order to ascertain what part is presently put under what uses and that are the prospects of its better use for the prosperity and well being of the population depending on it. The strategy should be that there remains no possibility for the misuse of the land resources. Stamp (1962) determines that land planning is an essence of the determination of the optimum use of every piece of land which must be elastic and can change from time to time to adopt the changing conditions.

The problem of landuse becomes more complex and intricate in mountainous environment where overwhelming proportion of the land is either not directly utilized or left unused due to adverse environmental conditions. The basic and most important use to fulfil the requirements is for agricultural activities for which hardly one third of the total land is presently available. Further, the productivity of land is restricted in many ways and is so poor that cannot be compared in any way with that of the conductive agricultural environment of plains. Moreover, the problem of soil erosion, land degradation, deteriorating ecological conditions etc. are the other important factors which focus the need for an instant conservation and reorganization of land in these areas.

Land degradation in Himalaya owes its origin, directly or indirectly, to the faulty landuse. Kumaun Himalaya is by not means an exception to this land degradation problem. There almost every patch of accessible land has been exploited. The problem of soil erosion is very alarming on account of very limited soil depth and steep slopes. The chances of recovery of soil according to experts are nil. The soil has reached the point of exhaustion and the economy has begun to collapse. These areas were a green and pleasant lands thirty years ago and today it is blasted health. Unscientific and indiscriminate deforestation and faulty land management have led to the miserable depletion of land resources. The uncontrolled flow of water from streams not only deteriorates the soil conditions but also creates havoc in plains through flooding and siltation. The agriculture is not in a state to fulfil the requirements of the region for more than five-six months in a year. The condition is turning from bad to worse day by day inspite of measures being taken by various agencies to improve the condition. This all emphasises the need for immediate land planning for its optimal use and management of these areas through detailed landuse survey of different parts. Since each region boasts of its own individuality, asserts its uncommon characteristics and agrarian and socio-economic problems, it requires a special treatment in its own way.

International Geographical Union has recommended that the landuse survey in India should be based on the British model advised by Stamp (1962). But this seems not practical in case of our country owing to certain limitations. The alternative suiting the landuse planning in India may be the modified form of Chinese method wherein the basis is the survey conducted by Buck (1937). It is confined to sample surveying of certain selected areas and the construction of a generalized picture of landuse.

Individual studies of different parts can contribute significantly in solving the local landuse problems. Shafi (1960) has emphasized the need for such local surveys to be conducted throughout the country for improving agricultural and social environment.

1.2. AN OVERVIEW OF LITERATURE

The idea of landuse studies is basically concerned with the ideas of British Geographers where in 1930 the landuse survey of every acre of land was recorded on maps. It was organized by Prof. Stamp (1948) in October 1930. The idea was to find out exactly what use was then made of every acre of land of Great Britain. The bulk of the country was actually surveyed and the land was recorded for the years 1931, 1932 and 1933. There remained small parts which had to be completed later but the whole was virtually finished before the out-break of war in 1939 and the plough-up campaign which made such an enormous difference of landuse in country (Stamp 1964).

Besides Britain, several other countries viz. U. S. A., Japan, China, U. S. S. R., Canada, Pakistan, Nepal, Cyprus etc. conducted landuse surveys. In U. S. A. the beginning of landuse survey can be traced from the work of Baker. He emphasized the need of land classification and surveys. U.S. Department of Agriculture also launched a programme of landuse survey in 1935 and 1938. The concept of American landuse planning as per view of Hill Man (1957), is as bread as country life itself and organisation therefore, presumably begun with the most general functional organization that can be discovered in an area.

In Japan the landuse surveys have been given the highest priority to support its huge population on the small area of cultivated land (Stamp 1964). In China sample landuse surveys were conducted by John Lessing Buck (Buck, 1937) and others highlighting the landuse problem in China. Landuse survey in many other countries are also in progress and being conducted by individuals as well as government and semi-government organization. In Nepal P. P. Karan (1960) had made a study of landuse map using air photographs.

Indian Geographers have long been attracted to study the problems of landuse in the country. In 1940, S. P. Chatterjee pointed out the necessity of undertaking landuse survey in India on the lines of that of Great Britain undertaken by Lord Dudly Stamp. Many Geographers and economists started landuse studies taking small and large units for study. S.P. Chatterjee (1945) proposed an atlas of land utilization in the District of Twenty Four Parganas of West Bengal. A detailed survey was also conducted in Hawrah and 1200 landuse maps were prepared. P. Dayal (1947) prepared a thesis on Agricultural Geography of Bihar. V. L. S. P. Rao (1947) made landuse study of Godavari Region and referred to the aims, objectives and classification of categories of landuse of the World Landuse Commission set up by the International Geographical Union. Mohd. Shafi (1960) emphasized the need of sample landuse survey and carried out landuse survey of Eastern Uttar Pradesh. Other eminent workers in the line who have carried out landuse surveys in different parts of the country and highlighted the various aspects of landuse planning, are U. Sen and M. Guha (1960), R. Lahiri (1950), S. B. Goswami (1960), A. Sharma (1962), R. N. P. Sinha (1960), K. N. Das (1956), R. L. Singh (1968), O. P. Bhardwaj (1961 & 1964), B. N. Ganguli (1955), U. Singh (1955), V. R. Singh (1950), B. K. Roy (1969), C. B. Joshi (1961), A. T. A. Learmonth (1962), K. Z. Amani (1965), B. N. Sinha (1948), D. S. Jalal (1970, 71, 76) Jasbir Singh (1971), V. M. Deshmukh (1975), M. Prasad (1978), B. N. Jha (1979) etc.

Besides, various other institutions and department like National Bureau of Soil Surveys and Landuse Planning ICAR, Agricultural Universities, Department of Science & Technology, Department of Environment, National Landuse and Wasteland Development Board, etc. are also engaged in carrying out sample surveys of various parts of the country.

The Himalayan areas have not received very much attention as far as the detailed landuse planning of the area is concerned. Studies carried out by S. L. Kayastha (1964), S. C. Shah (1973), Tewari (1966), Jalal (1970 and 71), Samant (1986), a few sample surveys carried out by Pantnagar Agricultural University and some studies by NIRD of Hyderabad are worth mentioning. The most important legendary work carried out in Kumaun Himalaya is by Jalal (1976) who has studied the landuse of a most remote part of the country i. e. the border district of Pithoragarh.

No landuse survey is however available for the Kotadun of Kumaun Himalaya which is the study area of present analysis. This part of the Kumaun Himalaya is an entirely different physico-cultural entity obviously distinguishable from the rest parts of Kumaun. The problems of landuse, the agricultural practices and other socio-economic and ecological problems are different owing to peculiar physical and socio-economic environment and area is most backward in reference to the Kumaun Himalaya as a whole. It calls for a specific landuse study and landuse planning. The agriculture is still of subsistence type. Activities outside agricultural sector are also not affluent to make the region self-sufficient. Ecological conditions are getting worse day by day with the indiscriminate cutting of trees and over exploitation of forest resources, soil depth and fertility is diminishing owing to the problem of soil erosion. Likely unwise and unscientific use of agricultural land as well as of the land under others uses creates innumerable problems. This all leads to a further deterioration in agriculture which is already in a very bad state. Keeping all these factors in view a landuse survey of this area is found to be immense and useful for the overall socio-economic and ecological development of the region. Moreover, such a case study can serve as a model for other hill areas or recommendation and findings of this study may be applicable to many of the environmental conditions similar to those of the area under survey.

1.3. CONCEPT

Land utilization is a comprehensive expression cannoting the use of land resource in such a manner that provides for a maximum exploitation of land area available. Planning and management of land refers to the conscious determination of use of land for the attainment of rationally desirable goals. Land has a geographical location and includes many components such as physical, biological, economic and institutional factors of the surface. It possesses variable capacity for different uses. The various uses of land can be studied in terms of land capability, land productivity, land efficiency, potential productivity, standard nutrition, land classification, land evaluation etc. It also involves the socio-economic conditions of inhabitants of an area since the use of any piece of land is very much determined by willingness as well as awareness of an individual or a community to put a land in its specific use. Moreover, in broadest sense, it includes all those man-made improvements which are attached to the surface of the earth and which cannot be separated from it.

In landuse planning there aught first to be the stage of survey—the recording of the present position; secondly, the stage of analysis—the seeking to understand the reasons for that position and of seeking what are the existing trends in development, and then, thirdly, the actual planning for future which must take the past and present trends into full consideration. The planner's task is not only to determine the optimal use of the piece of land but its multiple use whenever possible. The land planning must satisfy the basic needs of the inhabitants. There is a great need for integration between essential uses of land. Rival climents are each very apt to regard their particular requirement of land as being the most important. In doing so there are three golden rules to be observed. The first may be called the rule of optimal use, the second as the multiple use and the third that there ought not to be any waste land.

Landuse planning also aims at a much better adjustment of each category uses to the physical environment for the optimum utilization of land and conservation of natural resources. It also involves to propose a tentative scheme for the agricultural improvement based on the geographical potentialities under the limitations imposed by controlling factors. Landuse planning suggests the manner in which land resources are used and maintained for which group action is needed from time to time to discourage unwise wasteful practices, which may prove injurious to himself, to the neighbours and community and to society at large. Thus, the landuse study of a region in a synthetic form must bring out the personality to land-man activity, optimal and best use of each piece of land and the associate problems. The findings with their correct perspective should be a base for the formulation of an optimal landuse plan of the area.

1.4. The Study Area

Kotadun is the study area of present investigation. Extending in a general east-west direction, it is one of the five duns (Dehra, Kothari, Chaukhamba, Patli and Kota) in the Uttarakhand Himalaya. Along with its well demarcated boundaries of watersheds of Kunjkharak-Binayak range of Lesser Himalaya to the north and Babilia-Kailkhura range of Siwaliks to the south, Kotadun like other duns, is an elongated valley between river Kosi in the west and river Baur in the east. West of Nainital area, it extends between 29° 17′39" and 29° 30′40" N. lat., and 79° 9′13" and 79° 26′24" E long. (Figure 1.1) and encompasses an area of 360.78 km^2 and a population of 20657 persons in 1981. Including three forest localities, there are 96 inhabited and one un-inhabited villages.

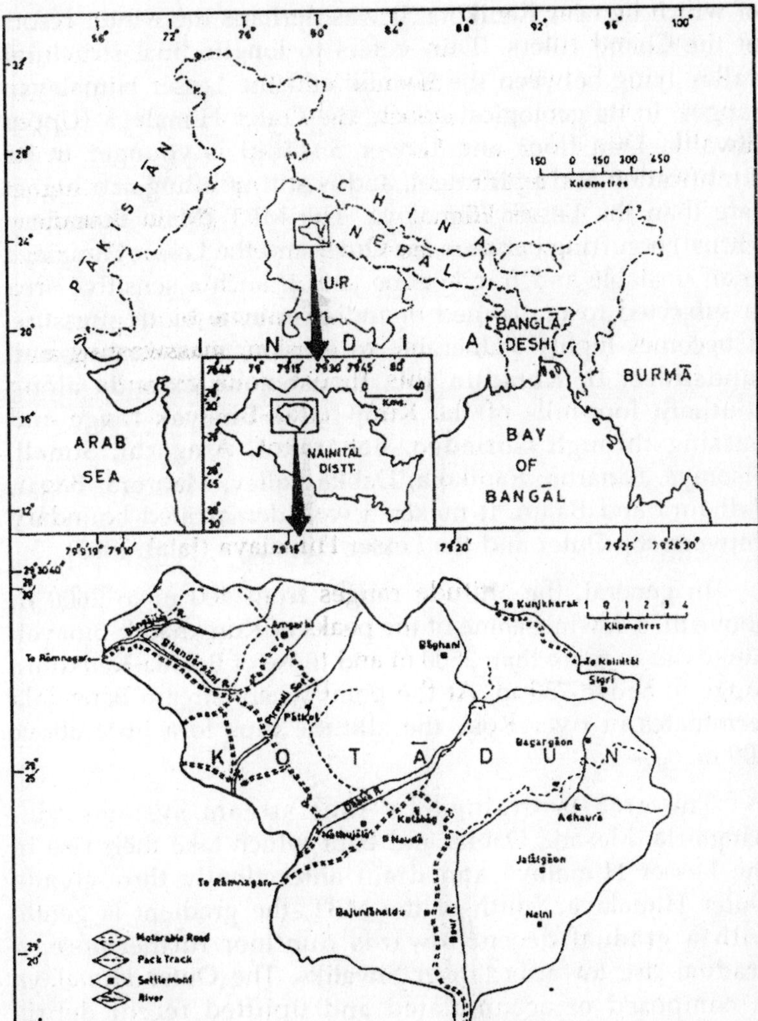

Fig. 1.1 Kotadun: Location Map

Kotadun is the combination of two words viz. Kota+Dun. 'Kota' refers to 'KOTHA' i.e. palace of Chand rulers the ruins of which lie near Ranikota. It was perhaps the winter resort of the Chand rulers. 'Dun' refers to longitudinal structural valley lying between the Siwalik and the Lesser Himalayan ranges. In its geological epoch, the Outer Himalaya (Upper Siwalik, Dun floor and Lower Siwalik) is younger in its stratification and upliftment, and is still uplifting at a higher rate than the Lesser Himalaya. The MBT (Main Boundary Thrust) occurring between the Outer and the Lesser Himalaya is an unstable and fragile zone, and if such a sensitive area is subjected to unplanned or indiscriminate biotic pressure, it becomes highly vulnerable to erosion, masswasting and landslides. In Kotadun this thrust zone extends along southern foot-hills of the Kunjkharak-Binayak range and passing through Goriadeo, Boharakot, Amgarhi, Simali, Sisoniya, Sanarha, Ranikota, Dabka valley, Mahrora, Bagar, Adhaura and Bajun, it makes a well demarcated boundary between the Outer and the Lesser Himalaya (Jalal, 1985).

In general, the altitude ranges from 400 m to 2600 m above m. s. l. while some of the peaks of Kunjkharak-Binayak range rise to more than 2500 m and those of Babilia-Kailkhura range to above 750 m. At the point where stream Bangajala terminates in river Kosi, the altitude dips to a little above 400 m.

The area is drained by four stream systems viz. Bangajala, Masani, Dabka and Baur which take their rise in the Lesser Himalaya and drain antecedently through the Outer Himalaya. South of the MBT, the gradient is gentle with a gradual decent towards dunfloor furthermore, a gradual rise towards Lower Siwaliks. The Outer Himalaya is composed of accumulated and uplifted recent debris carried down by the streams from the Lesser Himalaya. Coarse deposition comprising boulders, gravel and sand tends the streams passing through and dun, to drain subterraneanly during the lean season. The Outer Himalayan formation is suitable for extensively dense tropical-subtropical vegetation. North of the MBT, on the other hand, the gradient increases suddenly because of abrupt rise in the altitude. Steep slopes and crests of the Lesser Himalaya are uninhabitable and are covered with temperate vegetation.

Most of the settlements in the area occur near MBT on foot-hills or alluvial fans, in the vicinity of some perennial water source in the Lesser Himalaya. In the Dun floor, these have grown along the larger streams viz. Kakrar (feeder of Bangajala), Kaligad (feeder of Masani), Dabka and Baur, so that the water tapped at the foot-hills, is enough to meet the requirements.

The diversified topography, tectonic unstability and seismicity and the climatic variations (south facing slopes exposed to summer sun, high monsoon rains and cold winter conditions) combined with human interferences, have made this region fragile and hazards-prone. In the past three decades in the name of regional development for economic growth, more drastic damage has been done creating an ecological imbalance such as accelerated erosion, landsliding, masswasting, flooding, siltation, sedimentation etc., during monsoon rains and draughts in the dry months. Haphazard, indisciplined and short-sighted human activities such as ruthless exploitation of forest resources, overgrazing, unplanned cultivation, uncontrolled quarrying and constructions etc., consequent upon continuously increasing pressure of human as well as animal population, have caused severe land danudation affecting adversely the environmental conditions in the area under study.

The strategy of restoration of the damaged environment involves scientific investigations (land resource survey, land capability analysis and planning for optimal and multiple use of the land without deteriorating the eco-system) and intensive education to the society through demonstration and extension programmes at the sites affected. Present scheme is aimed at achieving these objectives in order to secure the ecological imperatives, utilize prudently the land capability, improve the economic condition of the society and restore a better equilibrium between the environment and the development of economic activities mainly the agriculture, forestry, grazing, mining and quarrying, settlements, roads and communications, industries etc.

1.5. AIMS AND OBJECTIVES

The project is aimed at:

(i) accomplishing a comprehensive landuse and land capability survey in Kotadun, Kumaun Himalaya to obtain information on the present landuse, its malpractices and consequent environmental implications,

(ii) formulating a scheme for optimal landuse taking into account the land capability, ecological imperatives and economic needs; and

(iii) educating local people in implementation of the optimal landuse scheme through demonstration and action programmes at two sites selected in Patkot and one in Saunjala.

The study thus includes:

1. drawing an overall picture of present landuse in Kotadun,
2. describing of unit wise landuse pattern of the area,
3. identifying of physical units,
4. explaining of physical and cultural environment of the area,
5. analysing of physical as well as cultural determinants of landuse,
6. explaining of landuse changes from 1954-55 to 1986-87 and cropping pattern of 1986-87 occurred in selected sample villages,
7. developing of model for land capability classification,

8. analysing of agricultural efficiency,
9. dividing of area in different altitudinal agro-climatic zones,
10. studying of land degradation with special reference to land slides,
11. suggesting of a suitable and feasible scheme for improving landuse and ecological conditions.

1.6. METHODOLOGY

The methods involved in carrying out the present investigation are as follows:

(i) Base maps are presented with the help of SOI toposheets specifically for relief, drainage, settlement, vegetation etc.

(ii) Maps for the sample villages are drawn from the revenue maps obtained from the tahsil headquarters of Nainital, Kaladhungi and Ramnagar.

(iii) Data pertaining to physical factors of relief, slope, drainage, forest and revenue area are based on the SOI toposheets through measurements by planimeter and rotarometer (map measurer).

(iv) Sample of soils and other have been collected during the field investigation.

(v) Climatic data for the sample sites collected with the help of the meteorological instruments set there and their day to day recordings. For other stations, these were collected from Forest Working Plans.

(vi) Primary data pertaining to landuse, population, livestock, physical as well as institutional infrastructure, cropping pattern, nutritional status and deficiency diseases etc. are collected through village level questionnaires from village Pradhans, elderly persons of the village, Patwaries, V.D.Os etc., while secondary data relating to these factors

have been collected from the Tahsil land record offices, Block and District Headquarters.

(vii) The area is divided into two main categories firstly, area owned by the State Forest Department and secondary, area owned by State Revenue Departments. Since the area owned by State Forest Department is managed and planned by the department itself, it is not taken into consideration in this investigation except the analysis of relief, soil, vegetation, geology etc. The area owned by State Revenue Department on the other land, is directly or indirectly under the control of the local society and is affected maximally it is therefore, taken into consideration for further details.

(viii) For detailed investigation three sample villages have been selected from the identified physical strata of dun floor, mid slopes and uplands.

(ix) The study area is divided into ten homogeneous physical units considered as basic units for the study. In delineating these units, the base maps pertaining to altitude, relative relief, slope, drainage and settlements are drawn separately. The physical location of each of the villages is identified by superimposing the settlement map over the drainage, altitude, relief, and slope maps turn by turn. Basin-wise location of all the villages are first observed and then these are grouped according to their altitudinal arrangements of two zones of below 800 m (dun-floor) and above 800 m (mountainous tract). Influences of other physical factors though very significant, are not taken into consideration in this particular case as a number of villages are likely to be divisible into several sections and it is not very feasible to go into these details. Each of the units thus identified, is named after the upper and lower sections of river basins (Table 1.1).

Table 1.1 Kotadun: Identified Study Units

Name of the basic unit		*Symbols used*
Bangajala	Upper	U_1
Bangajala	Lower	U_2
Mashani	Upper	U_3
Mashani	Lower	U_4
Dabka	Upper	U_5
Dabka	Lower	U_6
Baur	Upper	U_7
Baur	Lower	U_8
Others	Upper	U_9
Others	Lower	U_{10}

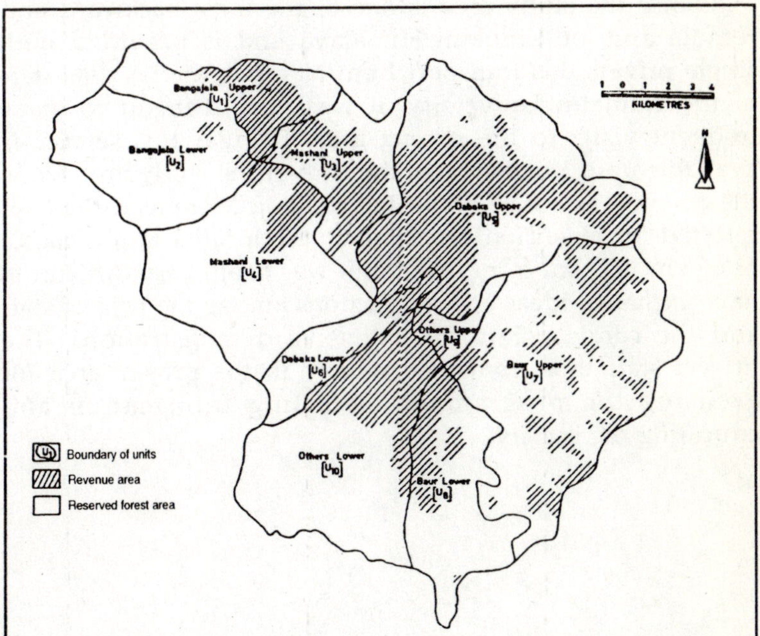

Fig. 1.2 Kotadun: Planning Units

(x) Three sites for action programme relating to optimal landuse, are selected in Kotadun, two in Patkot and one in Saunjala. Site-1 encompassing an area of 3.5 ha is the private land and technical advice is extended to the owner cultivators for implementation of optimal and multiple use of the land particularly for the purpose of agriculture, horticulture and farm forestry. Site-2 encompassing an area of 10 ha, is the public waste land which is selected exclusively for development of tree-culture including social forestry. Site-3 encompassing an area of 0.2 ha, is the private land. It is selected for development of farm forestry.

1.7. LIMITATIONS

Since the study area is one of the most backward and remote area of Kumaun Himalaya and is provided with single private bus transport from Nainital, the headquarters of the Kumaun University, it was very difficult to make frequent visits to the project area. Regular bus service is available upto Kotabagh i.e. the core of the study area while the country side is to be visited on-foot. Moreover, the sites selected for extension programme, are located at a distance of 66 km, 101 km from Nainital it was again very difficult to have exchange ideas and interactions among the project staff and the rural society for extensive demonstrations. The project staff thus, remained mostly in the project area for excusing the programme, collecting information and educating the society.

Chapter 2

Preliminary Report on Geology

The Main Boundary Fault (M. B. F.) divide the longitudinal structural valley of Kotadun into two litho-units of Lesser Himalaya to the north and Outer Himalaya to the south. The Lesser Himalayan domain is again subdivided into three formation groups of Krol, Blaini and Nagthat bearing mainly slate, dolomite, slate and quartzite. The Outer Himalayan zone is also subdivided into three belts of Lower Siwalik, Upper Siwalik and Dun deposition. The lithology on the Outer Himalayan barrier includes greywackes, conglomerates, slate and recent gravely strata. The Lesser Himalayan rocks dip 10° towards NWN while the Outer Himalayan rocks dip 22° towards SWS. Three sets of joint are observed in the Western section lying west of Dabka river. Highest joint frequency is observed towards NEE. Main sedimentary structures include conglomerate, ripple marks, crossbeddings and parallel lamination. In their geological epoch the conglomeratic boulders of the Upper Siwaliks formation are youngest one while the rocks of Blainy formation are oldest one.

2.1. LITHOLOGICAL SET-UP

After the works of Valdiya (1980) and Rupke and Sharma (1979) the lithological traits of the area are described by dividing it into two main litho-units:—

(i) The Lesser Himalaya, and

(ii) The Siwaliks.

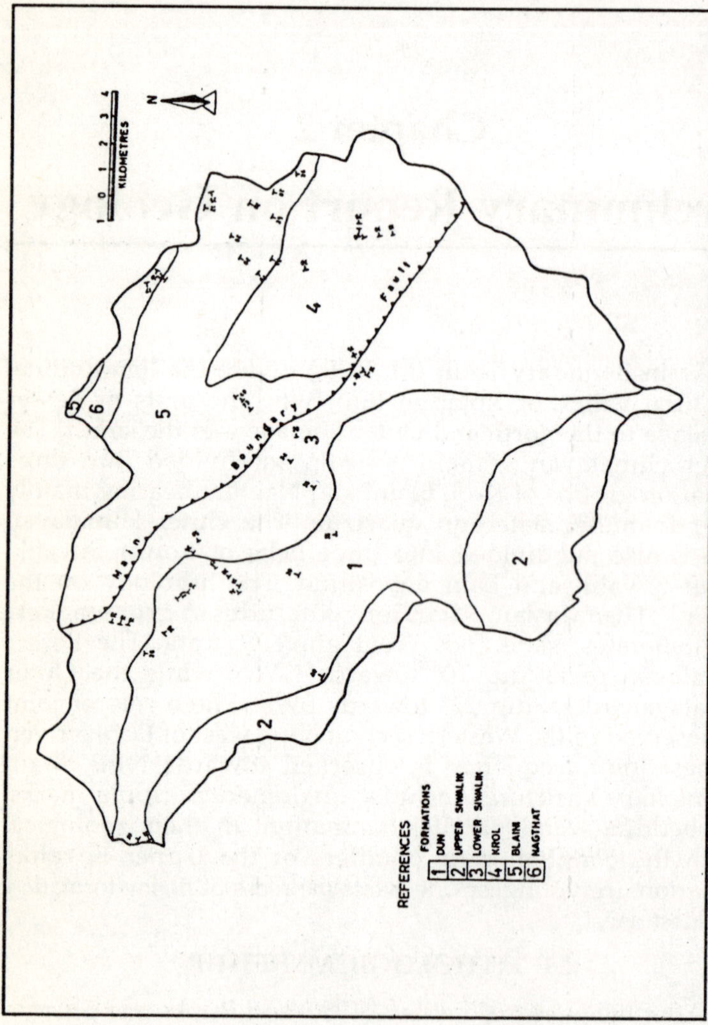

Fig. 2.1 Kotadun: Lithological Distribution

2.1.1. THE LESSER HIMALAYAN SYSTEM

The Lesser Himalayan domain lying in Kotadun may again be subdivided into three formation units of Nagthat, Blaini and Krol (Fig. 2.1).

The Nagthat Formations

The Nagthat Formations include purple faun and white and green quartzite which are of pebbly or conglomeratic in nature in some localities and are interbedded with greenish and purple slates. The word Nagthat is derived from the summit of Nagthat mountains situated at the Southeast of Chakrata. Auden (1934) has indicated a band of conglomerate between the Chandpur and the Nagthat Formations. However, the composition of this conglomeratic band cannot be an intra formation of local significance. The Nagthat formations are observed in Binayak-Kunjakharak range of Kotadun.

The Blaini Formations

Auden observed these formations in the valley of Blaini river lying in northwest of Solan in Himachal Pradesh. Blaini a corrupt word of Blaina, was introduced by Midilicott. The lithology of these formations includes conglomerate with admixture of siltstone, greywacke and slate of grey, olivegreen and black colours and impersisted with lenticular beds of purple or red limestone associated with purple slate and sandstone that rest upon the Nagthat quartzites in the Krol belt.

As per observations of Rupke, Gaur and Dave (1971) and Auden (1934) the Blaini is separated from the Nagthat by the sharpest inconformity of the Lesser Himalaya. Valdiya (1973 b) noticed a sharp lithological break in several localities of Dumaila, Sigri, Binayak and Pangot lying in the NE part of Kotadun.

The Krol Formations

Midilicott (1864) gave this name to the lithology observed by him in Krol Formation near Solan in Himachal Pradesh.

The lithology of this formation is considered of limestone of grey to greenish grey colour and purple slates, siltstones and massive dolomite in the upper strata. This formation is widespread in Bagar, Adhaura and Bajun area of Kotadun.

2.1.2. THE SIWALIK SYSTEM

The outermost hills lying between the Himalayan Front Fault (H. F. F.) to the south and the Main Boundary Thrust (M. B. T.) to the north fall in the Siwalik System of Kotadun. The name Siwalik is derived from Siwalik Hills near Haridwar where these formations were studied first. Similar rocks cbserved elsewhere in India, were given local names such as the Makran system in Baluchistan, the Manchar System in Sindh, the Tipam-Dupi-Tila and Diuhing system in Assam, the Irrawadi system in Burma etc. This group constitutes on extensive belt of cenozoic (middle miocene) formation. Fresh water sediments flanking the southern fringe of the area from the outermost range of the Himalayan mountain system. The sandstone, mudstone, shale and siltstone are the main rocks flanking over fluvial and interbedded flood plains in the area. As per observations based on the works of Rupke Kotadun contains all the three formations of Lower Siwalik, Upper Siwalik and Dun.

The Lower Siwalik

It is exposed along the Main Boundary Thrust and consists mainly of the steep dipping outcrops of shale and subordinate grey sandstone. Considerable variations observed laterally and vertically, occur in northern parts of the area towards northern limit which is delineated by the structural strip of the Main Boundary Fault.

The Upper Siwalik

The lithology of the Upper Siwalik is characterised by coarse boulders, conglomerates, thick earthy clays, sands and pebbly grit with admixture of old alluvium. The depth of these deposits is estimated from 1800 m. to 2750 m (Wadia 1981). The period of deposition is from Lower pliestocene to

Lower Pliocene. These formations are observed in the southern part of Kotadun.

The Dun Deposition

In order to observe the vertical cross sections of Dun deposits all the river courses of Kotadun were surveyed. Alternate tiers of pebbles, sand and clay are exposed for a depth of more than 60 m. along the river courses owing to deep down cuttings. The amount of dolomitic limestone and pebbles of quartzite is comparatively higher in the Baur valley. On account of chemical reactions on calcium, it is working as cementing material to compact these sediments. In other river valleys the deposits include higher amount of quartzite. Fine clay and sand are observed near Simli.

2.2 STRUCTURE

For the purpose of structural analysis the area is divided in two parts of Siwalik ranges lying east of the Dabka river and those lying west of the river. The weak structural planes (bedding and joints) were studied by taking the dip direction and strike line measurements with the help of Brunton Compass.

The ploted bedding planes of Western Siwalik ranges have the average trend is 17-197/10°→107° strike line, dip and dip direction respectively. Eastern Siwalik ranges have the average trend 22-202/22°→112° strike line, dip and dip direction respectively.

Maximum five set of joint pattern was shorted out which general trend of strike line, dip and dip direction is as 132-312/82→222, 268-88/50→178, 30-210/64→300, 29-209/40→119 and 78-258/50→348. Maximum joint frequency of western Dabka are dipping towards 140° to 150° direction while in the eastern Dabka these dip towards north.

2.3. SEDIMENTALOGICAL SETUP

Sedimentary deposits, not only record their pre-deposition history but these have imprinted on them some records of their depositional environment. The term environment has been much used but rarely given a precise definition. Krumblin and Sloss (1963) defined it as the complex of physical, chemical and biological condition under which a sediment is accumulated. Shepard and Moore (1955) defined it as spatial unit in which external physical, chemical and biological conditions and influences affecting the development of a sediment are sufficiently constant to form a characteristic deposit. Describing the importance of sedimentary structure Mckee (1971) believed it to be a more reliable guide to environmental analysis.

Sediments deposited by running water are classified in many ways. Most of the geomorphologists distinguish between vertical acceration and lateral acceration deposits. Lateral acceration occurs from the lateral migration of the channel, resulting in redistribution of the available sediments. Vertical acceration denoting vertical deposition and upward growth of the flood plain by deposition, is too simple, more over it has a drawback that even lateral acceration results in vertical ground and it is not always possible to distinguish the two.

External Bedding Forms

Some of the important aspects of the external bedding forms considered here include:

(a) Sand-Shale Bedding and

(b) Parallel Lamination.

The parallel lamination are noticed in white quartzite of the Nagthat and greywacke of Lower Siwalik formation. Shale is the compactness of small particles and such particles usually occur in the mouth of rivers. The sand which formed of coarse particles of rocks occur generally near the head of rivers. Parallel lamination is the indication of slow process of deposition.

Internal Bedding Forms

Crossbeddings: Valdiya (1964) and Valdiya and Bhatia (1970) have reported crossbedded units at various places around Nainital. Such type of formation is observed between two parallel beds. In Siwalik formation crossbeddings are frequently observed along river courses. According to environmentalists such beds come into existence because of sudden flood.

Gradded Beddings: In this type of bedding, coarse particles are associated with finer ones but show a continuous decrease in coarseness of sand particles. By such samples we can imagin about the environmental deposition process of the place where it had come into existence. According to environmentalists it is because of deposition in a geosyncline where continuous flooding occurs in varying speed. It is observed in quartzitic and greywackey rocks. A fine picture may be observed between Badhantoli and Binayak Range.

Bedding Plane Marking: Bedding plane structures of Kotadun may be divided into those on the base or sole of bed, one those on the top of the bed are more prominent. These include (a) Ripple marks and (b) Mud cracks.

Ripple marks are observed in Blaini and Nagthat formations. On the basis of these marks one may know the direction and velocity of current. Because of rare frequency it was not possible to assume the current direction. Mud crakes are observed in loose greywacke.

2.4. AGE OF THE ROCKS

Eminent scholars who divided the area in different formations also tried to estimate their age on the basis of the evidences occur in different geological times. These evidences are the fossils which are present in the deposits and the geo-carbon-dating process which in need for estimation of the age of sedimentary deposition.

On the basis of paleontological study Wadia (1981) estimated the age of the Lower Siwalik from Lower Pleistocene to Lower Pliocene, and that of the Upper Siwalik to post middle Miocene.

According to chronological sequence Krol is placed on the base of Tal Formation. Valdiya (1975) estimated the age of Krol formation to be younger than the carboniferous particularly the age of Upper Krol belongs to Upper carboniferous to Lower permian age. Auden (1937) placed these in the Permo-Triassic age on the basis of Blaini conglomerate which are related to Upper carboniferous. After the study of Nainital rocks Gansser (1964) concludes that if Nainital rocks are Krol, the rocks belong to Lower Paleozoic. On the basis of intensive field observations Valdiya confirms the Nainital rocks to be unquestionably Krol.

Blaini formation is studied by various scholars and various ages have been suggested for it. Holand (1908) places it in the time scale at Precambrian. According to Fuchs it belongs to Upper carboniferous. Valdiya further suggests that the Tal formation is of Permian period so that the Blaini cannot be young as Upper carboniferous.

Nagthat formation is suggested to be the production of Precambrian age (Valdiya 1961, 1964 & 1967) while Auden (1934 & 1937) showed it as a Lower Paleozoic probably Devonian formation.

Chapter 3

Morphometry

3.1. RELIEF ANALYSIS

The synclinal valley of Kotadun has a range of relative and absolute relief from 10 to 920 m. and 450 to 2623 m. respectively. It is surrounded by high ranges towards North and very low ridges towards South. Due to high gravity (height and slope) many environmental hazards are taking place. Hipsometric analysis indicates its erosional stage being between mature to near about young stage of its geomorphic orogeny.

Worcester (1948) describes the relief as the vertical irregularities both large and small of the surface of the lithosphere. Morphometric approach is the most popular one for the assessment and interpretation of these irregularities. It includes absolute and relative relief, profiles and hipsometric analysis.

3.1.1 Absolute Relief

The study area is divided into grids of one square kilometre. The maximum height of each grid is taken as the absolute relief marked by isopleth technique. S.O.I. topographical sheets $53\frac{0}{7}$, $53\frac{0}{2}$, $53\frac{0}{3}$, and $53\frac{0}{6}$ are used as the basic tools. The range of absolute relief is between 450m. and 2600 m (Fig. 3.1) increasing from south to north. The area is first divided into six zones having an internal of 400m. by isopleth technique. These zones are then grouped into three broad of below 900 m. 900-1800 m and 1800 to 2623 m for the present description (Table 3.1).

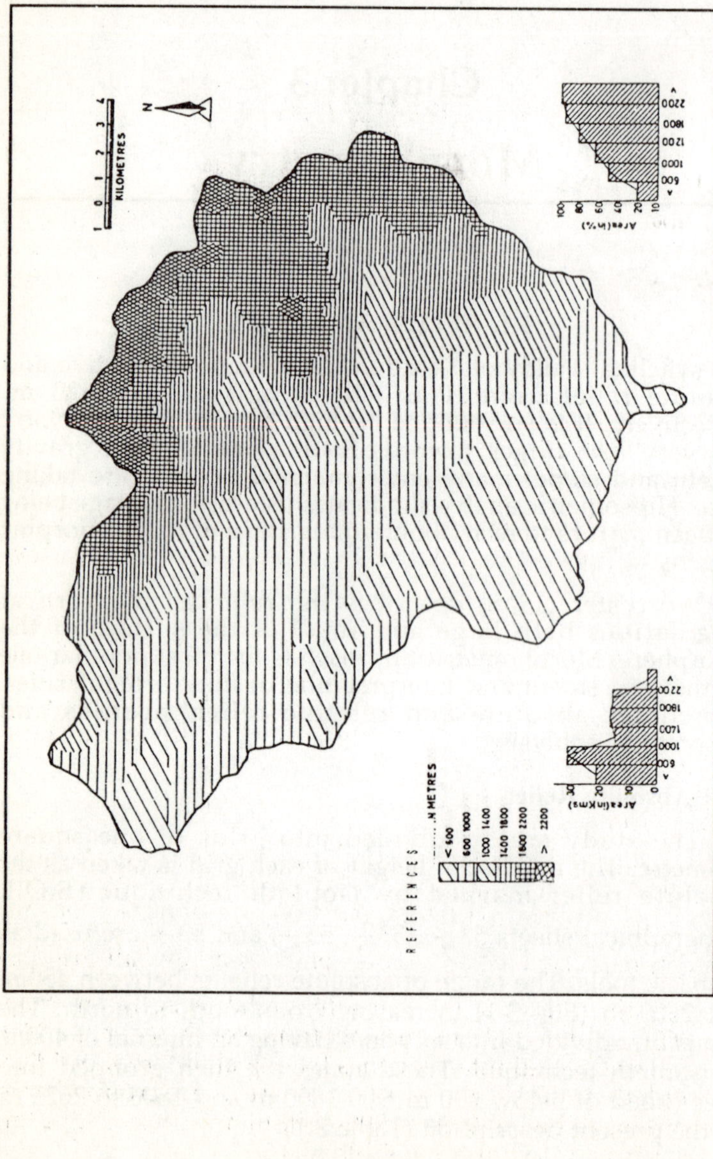

Fig. 3.1 Kotadun: Absolute Relief

Table 3.1 Kotadun: Absolute Relief

Height group (m)	*Area (Km²)*	*Cumulative Area (Km²)*	*Area (%)*	*Cumulative Area (%)*
< - 900	182.16	182.16	50.5	50.5
900 - 1800	109.99	292.15	30.4	80.9
> 1800	68.63	360.78	19.1	100.00

The zone lying below 900 m. above m. s. l. occupies 182.16 km^2 or 50.5 per cent area. It entirely lies within the dun floor. This zone is generally level with uniform ground except in the case of river courses bearing U shaped gorges owing to rapid down cuttings.

The area lying between 900-1800 m. covers 109.99 km^2 or 30.4 per cent area. Major proportion of this zone is occupied by Siwalik formations and a small one lies under Lesser Himalayan ranges. This is highly dissected zone owing to inter actions of orogenetic and erosional forces creating variety of land forms of sharp peaks, steep scarps, hillocks and gorges.

The 1800 to 2600 m. elevation zone comprises 68.63 km^2 or 19.1 per cent of the total area of Kotadun and occurs in the north and north eastern part of it. This zone includes bare rock scarps.

3.1.2. Relative Relief

Relative relief represents the difference of maximum and minimum heights of a unit. It is computed by placing the difference values in 1 km^2 grids. Using these values Kotadun is divided into six isopleth groups having an interval of 125 m. For further description these are taken into four major groups (Fig. 3.2) of 0-125 m, 125 to 375 m, 325-500 m and above 500 m. (Table 3.2).

Table-3.2 Kotadun: Relative Relief

Height Group (m)	*Area (km^2)*	*Cumulative Area*	*Area (%)*	*Cumulative frequency*
<-125	138.84	138.84	38.48	38.48
125-250	32.656	171.496	9.06	47.54
250-375	33.532	205.028	9.29	56.83
375-500	91.098	296.126	25.25	82.08
500-625	47.855	343.981	13.26	95.34
>625	16.799	360.78	4.66	100.00
Total	360.78	-	100.00	-

First group of 0-125 m includes 38.84 per cent of the total area under which there lie the entire Dun floor and the small hillocks of Siwalik ranges. So, the isopleth of 125 m. relative relief is far away from the South. The physical structure of this zone is very simple. The Dun floor is almost plain but the river courses are very deep and create undulation in the surface.

Second relative relief group having 18.35 per cent area lies between 125-375 m. Above 125 m. the isopleths of 250 and 375 m. come close to each other. The height increases rapidly towards north so the isopleths become closer. As the height rapidly increases, the area becomes smaller. The old landslides observed in Lower Siwalik ranges seem to have occurred because of the sudden height variations in this zone.

Third group of 375 to 500 m. relative includes 25.25 per cent area. Above 375 m. the isopleths stop their closeness owing to increase in area along with increasing order of height. This zone is occupied by dense human habitats. Geologically there lie the Lower Siwalik and the intermittent zone of Siwalik and Lesser Himalayan ranges. Along the Main boundary Fault this zone is disturbed by endogenetic forces indicating its successive dislocations resulting in environmental hazards particularly those of landslides.

Fourth group of above 500 m. height includes and area of 17.92 per cent. There are higher ranges of Lesser

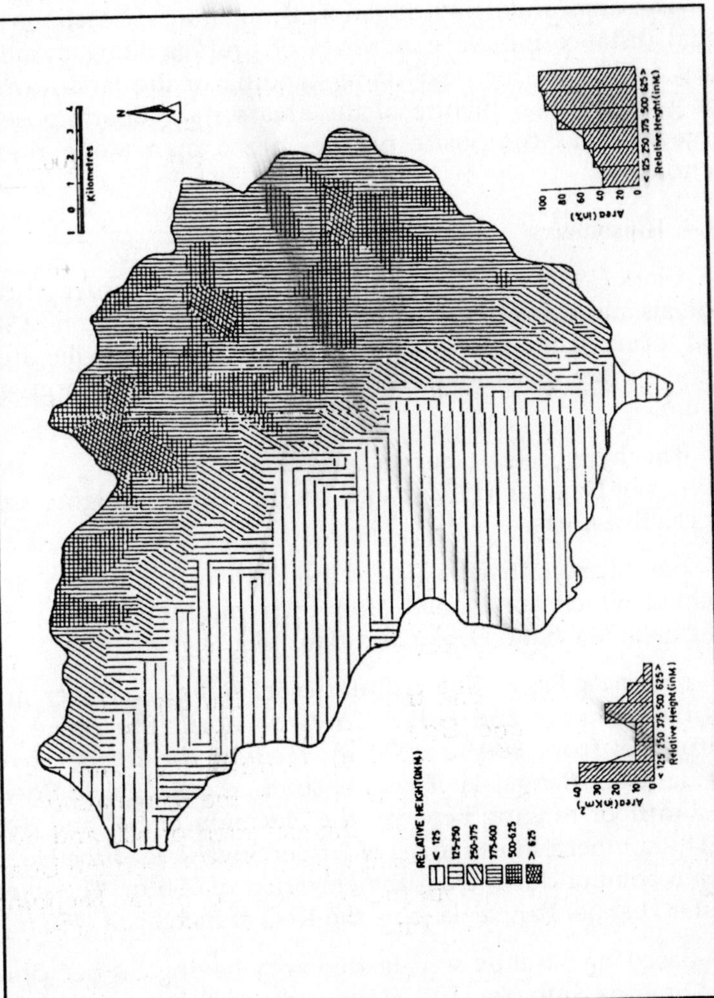

Fig. 3.2 Kotadun: Relative Relief

Himalayan system. So the contours come close to each other decreasing the area lying there in.

3.1.3. Profiles

The area under investigation is cut into six sections of equal distance intervals. A series of profiles along parallel lines are drown for clear understanding of the land-forms. For getting a real picture of the area serial, superimposed, projected and composite profiles are drown along these sections.

3.1.4. Hipsometry

Clark (1966) defined the hipsometry as the mathematical analysis and measurement of the shape and dimension of its land-forms. It is used to indicate the proportion of the area of the surface at various elevations above or below a given datum.

The hypsometric curves are drawn according to the method of Strahlar (1952) based on the relative heights and the relative areas.

For hipsometric analysis the fifth order basins are selected which are named after their master streams viz. Bangajala, Mashani, Dabka and Baur (Fig. 3.4).

Bangajala Basin: The tributaries of Bangajala river are Kotilia, Kakurar and Udar. Towards north the range of altitude is from 800 to 2000 m. feeding the former two tributaries of Bangajala. Towards south, the Bangajala flows on dun floor ranging between the elevation of 600 and 800 m. The southern-most ranges of Upper Siwalik feed the Udar stream coming down from the elevation of 750 m. The joint master channel Bangajala joins the Kosi at Bangajala (450 m).

Based on Strahter's scale this area having 32 per cent hypsometric integral, lies under the mature stage of its geomorphic development.

Mashani Basin: It comprises an area of 72.13 km^2. The height range is between 450 m and 2200 m. above m. s. l.

The hipsometric integral of 46.6 per cent indicates the mature stage of geomorphic development of Masani Basin.

Dabka Basin: Dabka the largest river in Kotadun, rises on southern slopes of the outermost ranges of Lesser Himalaya culminating above 2600 m above m. s. l. Southwards it flows through the Dun floor descending to 450 m. above m. s. l. The basin occupies an area of 82.5 km^2. Hipsometric analysis of the basin indicates that it has reached to the stage of maturity of its geomorphic development. But the hipsometric integral of 56 per cent indicates its youthful stage.

Baur Basin: Encompassing an area of 100.25 km^2, the Baur basin is the largest one in Kotadun. The maximum length of river is 20 km. The maximum height of the basin is 2122 m. and the lowest one is 450 m. Hipsometric integral of 51.2 per cent puts in into mature stage of its geomorphic development.

Hipsometric integral indicates remarkable results striking the attention towards opposite relation of the erosional surface. The unique feature of Himalaya comes into existence due to recent geomorphic process being the product of Lower Pliestocene to post middle Miocene. Although the Hipsometric analysis puts it under the mature to near about young stage of erosion but the dun orogeny is youngest one in the Himalayan domain. The researchers consider it because of its structural formation.

3.2. SLOPE ANALYSIS

Slope is defined as the angular inclination of terrain between the mountain tops and the valley bottoms resulting from a combination of many causative factors like structure, climate, vegetal cover, drainage texture and frequency, dissection index, relative relief etc. which are significant geomorphic attributes in the study of land-forms of drainage basin (Calef, 1950). In Himalayan region generally the slope is controlled by the bedding dips, but the erosionary process is active to deform its inclination dominant in several places which is subjected for creating environmental hazards of soil erosion, mudflow and landslide.

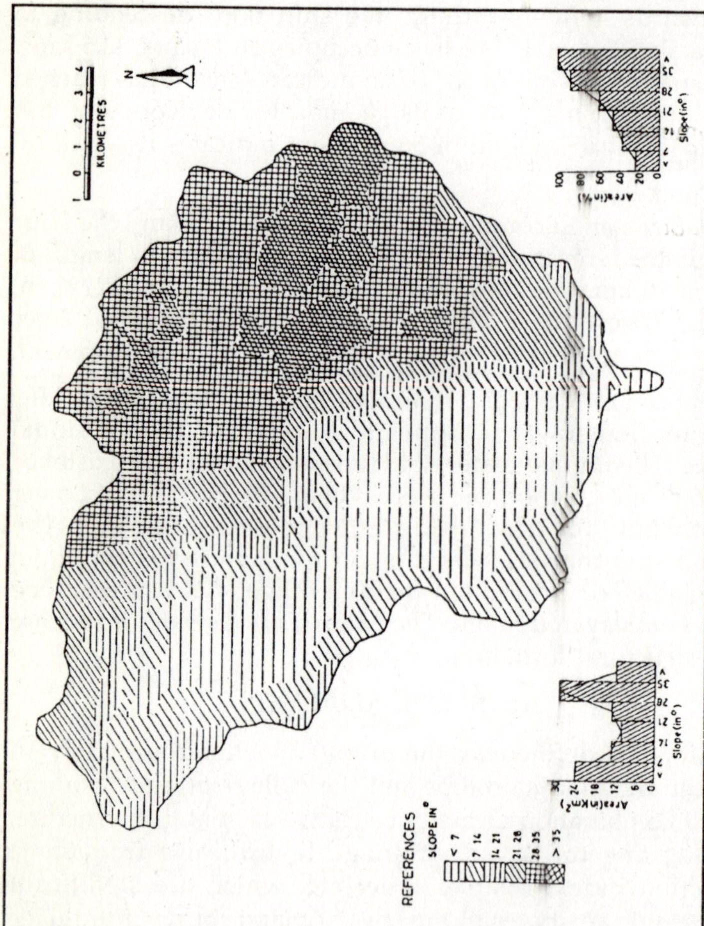

Fig. 3.3 Kotadun: Average Slope

In order to understand the nature of these hazards, it is necessary to understand the morphology of an area. In Kotadun the slope is dominant causative factor. The aim of the study is to divide the area into various slope categories and to assess them statistically in various ways viz. central tendency, dispersion and correlation with absolute and relative relief.

The slope analysis is prevalent since the very beginning of geomorphological studies. A remarkable attention towards its scientific study is given by Rich (1916), Wentworth (1930), Raize and Henry (1937), Smith (1938), Robinson (1948), Calef (1950), Strahler (1956), Miller (1960) and Eyles (1965). On the basis of the method of Wentworth. Singh *et al.* (1977) divided the area into various slope categories. Present study of slope analysis is made by method of Wentworth which is modified in matric system. SOI Topographical sheet Nos. 53°/7, 53°/2, 53°/3, 53°/6 have been the basic tool for calculating the degree of slope. The slope is calculated with the help of the formula introduced by Wentworth (1930) modified in metric system.

$$Q = \tan^{-1} = \frac{N \times CI}{636.3}$$

Regional Distribution of The Average Slope

The average slope for each grid of 1 km^2 was calculated with the help of the formula given by Wentworth and the isopleths were drawn by making five slope category ranging from below 7° to about 35° with an interval of 7° (Fig. 3.3).

The distribution of the average slope in the area presents a large variation from region to region but it may be said in general that level slope region occurs in the dun floor while towards the south and north there occur the high slope regions.

Level Slope Zone: Those parts occurring below 7° are included in this slope zone. This zone covers an area of about 86.18 km^2 which is 23.89 per cent of the total area. Generally this zone lies in dun floor and a level area between small ranges of Siwalik hills.

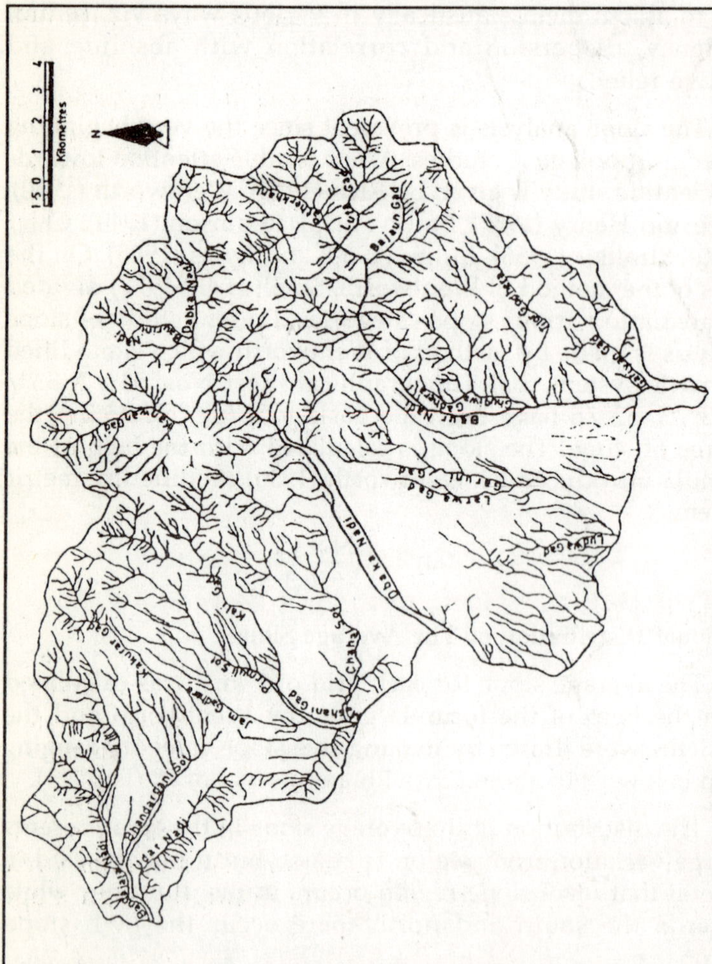

Fig. 3.4 Kotadun: Drainage Network

Table 3.3 Kotadun: Average Slope

Slope (0°)	*Area in Km²*	*Area%*	*Cumulative frequency (%)*	*Slope category*
0-7	86.180	23.89	23.89	Level
7-14	48.925	13.01	36.90	Gentle
14-21	36.075	10.00	46.90	Moderate
21-35	149.575	41.45	88.35	Fairly steep
> 35	42.025	11.45	100.00	Steep
Total	360.78	100.00	—	

Gentle Slope Zone: This zone contains 46.92 km² or 13.01 per cent of total study area. It includes Upper Siwalik ranges and foot hills of Lower Siwalik. Just above the dun floor there happens a sudden change of uneven surface but the low ranges of Siwalik subjected to create such gentle slope.

Moderate Slope Zone: The Upper Siwaliks and the foothill zone of Lesser Himalaya between the average inclination of 14-21° covering an area of 30.075 km² is included in this slope zone.

Fairly Steep Slope Zone: Covering an area of 149.57 km² (41.45%) this includes the area where the average slope ranges between 21-35° except the summits of Lesser Himalaya where the relative relief is high and rocks are comparatively hard.

Steep Slope Zone: This slope zone includes those parts where the average slope is between 35° to 42°. Covering an area of 42.025 km² (11.65%) this zone extends in the summits and their surroundings.

Central Tendencies & Dispersions

The average slope in Kotadun ranges from 7° to 42°. The mean, median and mode of the average slope values stand at 28.87, 22.68 and 22.28° respectively. A measure of central tendency tells something about the general level of magnitude of the distribution. It does not tell how the items scatter around the measure of central tendency.

Average values give us single value of series and conceals the variation among the values. Measure of variation or dispersion explains how the items are dispersed away from the average Dispersion includes the deviation, inter quartile range skewness and kurtosis (Table 3.4).

Table 3.4 Kotadun: Statistical measures of the Slope

Mean	: 28.87		
Median	: 22.68		
Mode	: 31.36		
Deviation:—			
	From mean	–	11.41
	From Median	–	11.28
	From Mode	–	12.52
$Quartile_1$		–	7.60
$Quartile_3$		–	10.93
Quartile D.		–	1.67
Skewness		–	0.31125
Kurtosis		–	2.17

Table 3.5 Slope Vs Absolute Relief

Metre/Degree	0-7	7-14	14-21	21-35	>35	Total
0-600	44.505	25.22	6.13	–	–	75.855
600-1000	41.675	21.705	21.715	16.894	–	101.989
1000-1400	–	–	8.23	37.304	6.225	51.759
1400-1800	–	–	–	46.089	12.15	58.239
1800-2200	–	–	–	34.819	20.35	55.169
>2200	–	–	–	14.469	3.3	17.769
Total	86.18	46.925	36.075	149.575	42.025	360.78

Table-3.5 reveals that with an increase in slope absolute relief also increases suddenly and when absolute relief is very high slope inclination also reach very high. Where slope is above 35° absolute relief is 1400 m to 2623 m. and where slope is between 0-7°, absolute relief is also 0-1000 m. In this way it is clear that the slope rises along with the rise in altitude. The correlation coefficient is calculated according the Karl Pearson method. The correlation value +.82 express as well as absolute height increasing slope also increases.

Table 3.6 Kotadun: Slope Vs Relative Relief

Metre/Degree	0-7	7-14	14-21	21-35	35	Total
0-125	96.1214	36.663	9.3809	3.6747	–	
125-250	.0586	7.262	21.4992	3.8362	–	
250-375	–	–	5.1949	26.8049	1.5322	
375-500	–	–	–	70.0812	21.0168	
500-625	–	–	–	34.341	13.514	
>625	–	–	–	10.837	5.962	
Total	86.18	96.925	36.375	149.575	42.025	

According to table -3.6 under 0-7^{0} slope include 0-125 relative relief and steep slope category include 250 to 625 m. relative relief. This table also shows when relief rise average slope degree also increase. The correlation value is +.84 and express as well as relative relief increases slope is also increasing.

3.3 DRAINAGE CHARACTERISTICS

In the field of geomorphology although there are many factors continuing from the origin of the earth but as far the erosional land-forms are concerned rivers are considered the most dynamic factor. The main rivers of Kotadun are Dabka, Baur, Masani and Bangajala. These rivers flow from an elevation of 2623 m. to 450 m. coming down from high degree slopes of Lesser Himalayan ranges these rivers flow across the monotonous topography of dun floor. Thenceforth these make deep gorges at the cuttings of Upper Siwalik. Almost all the rivers are seasonal in the dun floor because the loose lithology leads the water to flow subterraneanly. Main drainage patterns are of dendritic, parallel, trellis and intermittent types.

Outwashing the land material and changing the landscape the dry channels in dun floor begin to roar from early July for a few hours. Meandering over the recent pebbly strata of dun floor the flood water added with resurfacing of the sub-terranean water through flows the deep gorges of Siwalik ranges. The spring fed nature of rivers in the Lesser Himalaya attracts for the deep understanding of drainage characteristics. Geomorphic aspects of the study include the system pattern and density of the drainage.

3.3.1 Drainage System

The Baur System: Maximum length of the river is 20 km. originating from an elevation 2122 m, it is joined by many streams—Patherkhani from 2430 m. and Lura from 2116 m. Garrain, Lura and Bajun collect their water along the Main Boundary Fault. The lowest point in this basin is about 450 m. above m. s. l. Lying near the eastern boundary of the area the catchment covers the largest area of 100.25 km^2.

The Dabka System: Maximum length of the river is 20 km and the catchment comprises an area of 82.55 km^2. The river divides Kotadun into two equal parts. Torrents flowing in the Lesser Himalaya make it destructive owing to the occurrence of huge boulders and washing out the surfacial part in rainy season. Downstream in the dun floor, it forms a fertile terrace near Kotabagh. It has two head-waters—the eastern channel rises at an elevation of 2496 m. and the western one at 2478 m. above m. s. l. In dun floor its coarse is about 1 km. wide.

The Bangajala System: Maximum length of this river is about 15 km. It comprises a basin area of 51.93 km^2. The main streams joining the Bangajala river include Jhulwar, originating from 1400 m, Kakurar from 2180 m, and Udar from 702 m.

The Mashani System: Comprising a basin area of 72.13 km^2 its maximum length is 14 km. The maximum height of this basin is 2542 m. at the place from where the Kali the main head water channel, takes its origin. Other tributaries are Pirmuli Chahal and Jam Gadhera.

Remaining part of Kotadun is drained by myriods of small streams and because of plain surface, these use to flow towards the Bhabar. Except in rainy season, these remain in dry courses. The main streams are Sanani Ladwa and Gurrani. (Fig. 2.4)

3.3.2. Drainage Pattern

Seasonal and Perennial Rivers: There is direct impact of lithology on the drainage system. Hard rocks prevent percolation of water whereas the loose material accepts it. Seasonal rivers generally occur in the Siwalik ranges and the dun floor consisting loose material such as pebbles, sand stone, conglomerates, slate etc. The Lesser Himalaya consisted of hard rocks of quartzite, dolomite and shale provide good condition for the storage of underground water representing its manifestation in the form of seepages (Naula) and springs (dhara) giving a perennial water flow to the Dabka, Kakurar, Kali and Baur rivers.

Consequent and Subsequent Rivers: Before the emergence of Siwalik ranges there was a sea basin sedimented by many rivers coming down the Lesser Himalaya. During the course of upliftment of the Siwalik ranges these rivers maintained their channels through increasing the rate of course erosion. Dabka, Baur, Kakurar and Kali following their original courses, provide fine examples of anticedent drainage pattern. The streams came into existence after the upliftment of siwalik hills provide the consequent drainage pattern. These streams include Gurrani, Tilor, Udar, Lurwa etc.

Dendritic Pattern: This pattern, woven by head streams, exists generally in the watersheds of the Lesser Himalaya and the Siwaliks. Dendritic pattern of drainage is only as a net of first, second and third order streams. In the Siwalik region these streams are rainfed while in the Lesser Himalaya these are both rainfed and springfed. Best examples of this pattern are observed in between 1600 m. and 2400 m. in Lesser Himalayan region and 600 to 750 m. in the Upper Siwalik. In the eastern side, these streams are led by Main Boundary Fault in other areas a considerable part is covered by the activity of gravity.

Parallel Drainage Pattern: The streams rising from Siwalik ranges flow in almost parallel to each other. In dun floor generally each river flows parallel to each other. In Siwalik towards western side, first order river of Udar and Kali make this pattern which are mainly affected by gravity and loose lithology and woven by first and second order rivers. Fifth and fourth order rivers coming down from northern range, enter in dun floor and flow near about in a parallel pattern. These are Kakurar, Pirmuli, Kali, Dabka and Baur.

The Trellis Pattern: This drainage network has been observed around eastern side of Kotadun which is drained by Dabka and Baur systems. Contributors of water for Baur system including Patherkhani and Bajun streams along the M. B. T., form such patterns. This pattern is evolved in the zone of M. B. T. in the Baur and Dabka system due to the abrupt change in its course. Following nearly parallel to M. B. T. Bajun joins the Patherkhani at about 80°.

Intermittent Drainage Pattern: The area fed with gravelly material by the seasonal drains, is provided with intermittent drainage network where the torrents contribute the water to their consequents. In this network, the streams flow beneath the gravelly strata due to the absence of impervious rock bases. The intermittent pattern is observed in the dun floor. Kakurar, Kali, Dabka and Baur rivers are the examples of this pattern. The lithology is the composition of gravelly strata which is effective condition for this pattern.

3.3.3. Drainage Density

Horton (1932) defines the drainage density as the length of stream per unit area. He formulated as:

$$Dd = \frac{L}{Ab}$$

Where Dd is the Drainage density L is the total stream length and Ab is the area of the basin.

The drainage density is calculated on the basis of the measurements of rivers in the grids of 1 km^2 SOI topographical sheets. The results of these measurements have been transformed in isopleth map.

Low drainage density group contains 80.32 km^2 which is 22.26% of the total area. The maximum part of this group is below 7° average slope which is almost level area. The loose lithology and level ground of dun floor causes very high porosity owing to which the drainage network development is very poor. The drains collecting water from higher slopes are responsible for the drainage density pattern of dun floor having single drain.

Moderate density group contains 164.7 km^2 which is 45.65% of Kotadun. The area under this density group covers the range of average slope between 14° to 35°. The maximum part of Lesser Himalayan region covers this group.

High drainage density group includes 43 km^2 which is 29 per cent. The maximum part is under 28 to 35° average slope. The cause of high drainage density is weak geology of Siwaliks and the summits Lesser Himalaya.

Chapter 4
Climate and Natural Vegetation

4.1. CLIMATE

There are wide variations in the climatic conditions of the area. These are totally influenced by the relief. In southernmost part maximum temperature of 40° cg. is observed in the month of June while it is 15° cg. towards northernmost one. The minimum temperature of winter season reaches below 0° cg. along Binayak Kunjakharak zone. But there is not much variation in the distribution of rainfall.

There is no systematic presence of meteorological stations in study area. On the basis of the available records of surrounding areas it has been tried to assume the climatic conditions of Kotadun resembling the same conditions of the area. Considering similar physical conditions, the records of Manora peak are presumed for Pangot, Patwadanger for Nalni and Kaladhungi for Bajunia Haldu. Climatic records for Leti and Patkot are based on the records maintained in the Forest Range office Leti and the Site office of Kotadun Optimal Landuse Research Project, Patkot.

4.1.1 Temperature

Each of the plant species has an optimal temperature requirement associated with its functions of photosynthesis,

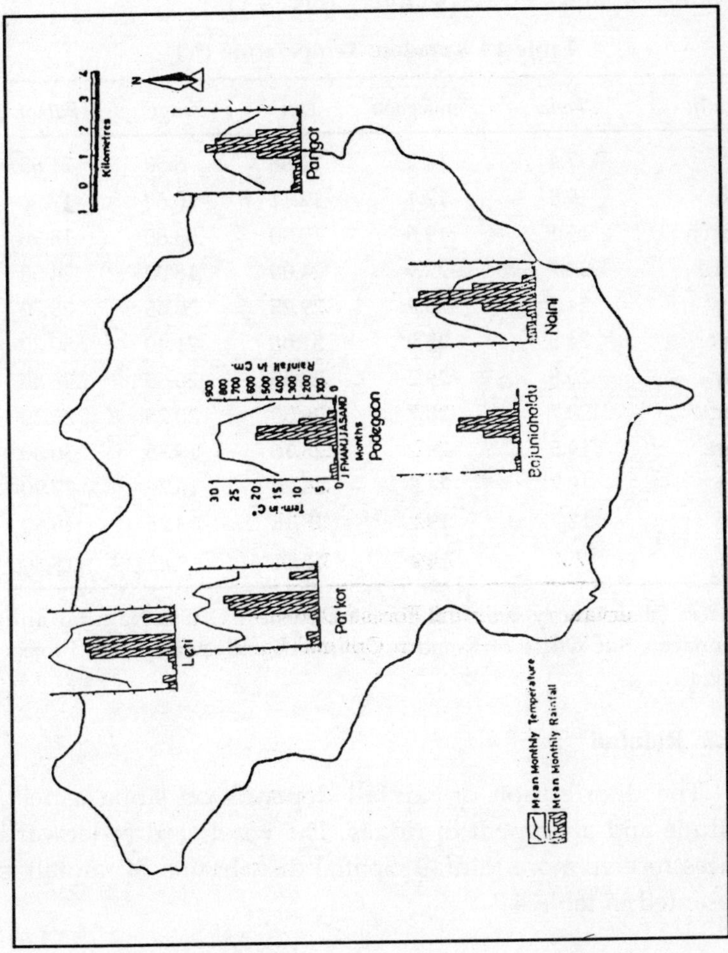

Fig. 4.1 Kotadun: Temperature and Rainfall

flowering, fruiting, and seed germination. In the study area rise in elevation gives decrease to temperature and hence, higher altitudes get the lower temperature. It is clear from table-4.1 that in the month of June, the highest temperature of Pangot is 21.40° cg. it is 24.5, 28.9 and 31.0° cg. at Nalni, Pandegaon and Leti respectively (Fig. 4.1).

Table 4.1 Kotadun: Temperature (°c)

Month	*Nalni*	*Pandegaon*	*Leti*	*Pangot*	*Patkot*
Jan.	7.4	14.2	11.08	8.50	21.65
Feb.	9.3	17.1	12.11	10.60	17.60
March	15.1	19.6	19.90	13.80	18.90
April	20.7	27.9	24.00	18.00	24.60
May	24.2	28.9	29.25	20.85	28.70
June	24.5	28.9	31.00	21.40	30.20
July	20.8	29.2	28.55	20.55	25.80
Aug.	20.7	28.7	26.30	20.25	26.20
Sept.	19.5	28.2	25.16	19.35	30.30
Oct.	18.7	24.3	24.70	16.70	27.90
Nov.	12.7	19.2	18.16	13.25	18.90
Dec.	7.7	14.4	12.11	9.85	18.60

Source: Observatory Nainital Forest Division Office Nainital and Ramnagar, Site Office of Kotadun Optimal Landuse Research Project Patkot.

4.1.2. Rainfall

The distribution of rainfall depends on slope aspect, altitude and alignment of ridges. The windward or leeward places receive more rainfall. Spatial distribution of rainfall is presented in table 4.2

Table 4.2 Kotadun: Rainfall (mm)

Month	*Nalni*	*Pandegaon*	*Leti*	*Pangot*	*Patkot*	*Bajunia-Haldu*
Jan.	58.41	38.7	58.25	76.8	51.50	30.6
Feb.	59.96	55.0	97.50	53.0	89.00	34.7
March	62.98	23.8	13.75	60.5	14.65	9.1
April	26.13	11.5	21.00	44.4	19.00	10.3
May	67.10	30.7	51.75	67.8	50.70	51.0
June	387.57	230.8	622.00	320.1	613.00	266.6
July	881.84	585.9	660.50	689.3	649.00	463.0
Aug.	847.40	578.8	667.00	612.2	675.00	446.7
Sept.	461.31	270.2	–	328.4	–	239.1
Oct.	92.16	61.4	–	24.7	–	36.5
Nov.	9.72	8.7	220.00	12.7	208.00	3.7
Dec.	24.43	18.8	26.50	29.7	21.00	2.5

Source: Observatory Nainital Forest Division Office Nainital and Ramnagar, Site Office of Kotadun Optimal Landuse Research Project Patkot.

4.1.3. Seasonal Rhythm

Annual variation of temperature and rainfall causes annual seasonal rhythm, Kendrew (1937) has classified the seasonal rhyth m as follows:

1. N. W. Monsoon Season From December to February.
2. Hot Weather Season from March to June.
3. S. W. Monsoon Season from June to September.
4. Retreating Monsoon from October to November.

1. The Cold Weather or N. W. Monsoon Season

The cold weather season (*Hyun*) is from beginning of December to end of February. In this duration the weather conditions are fair sky, low temperature and humidity, slow blowing northern wind etc. Fair weather condition is suddenly disturbed as a overcast by SE branch of the mediterranean depressions divided by the high air pressure belt of the Turkey. As a result there occurs a little rainfall towards valleys and snowfall in the area of Kunjakharak and Binayak range.

2. The Hot Weather Season

Form the beginning of March to last of June a sudden change comes in the weather of the study area. Locally this season is Known as *'Ruri'*. The sky remains almost cloudless but on the arrival of cyclones in the month of March followed by slight rainfall and hailing. Other remaining months are mostly rainless but in the mid of June two or three tarbulent storms followed by hails and thunder are common.

3. The S. W. Monsoon Season

It begins from the end of June to the end of September and is known as *'Chaumas'*. In this duration the advent of the Monsoon brings a complete change in weather and results in heavy rainfall. Hot sun is relaxed by it for the people. With the commencement of the monsoon which reaches here upto the end of June, the temperature goes on decreasing. However, break of rainfall causes sudden rise in the temperature. Generally the normal temperature and high relative humidity are the main characteristics of this season.

About 85 per cent of the total rainfall of the study area occurs in this season. After September humidity goes on decreasing and due to direct outgoing radiation, dew is most common in the morning.

4. The Retreating S. W. Monsoon Season

It begins from the beginning of October to the end of November. The temperature begins to decrease by the end of September. But durinal range of temperature decrease vice versa. In the early October the retreating monsoon drops a little rain followed by hails. The high durinal range of temperature causes the dew everywhere.

4.2. NATURAL VEGETATION

High altitudinal variation from 450 to 2623 m. indicate various differences in spatial distribution of vegetative cover. Below 800 m. main vegetation is Sal (*Shorea robusta).* Moist mixed decidous forest cover is observed along perennial water sources. Above 1200 m. the main species are blue pine (*Pinus excelsa*). Dry mixed decidous forest occur between the sal and pine forest.

As per suggestions of environmentalists the green vegetative cover is the best measure for the treatment of environmental hazards in Himalaya. Field observations reveal that frequency of the environmental hazards increased during the last 10 years. On account of rapid removal of protective green cover the tectonic activities become more active. Main characteristics of natural vegetation of the area are based on Forest Working Plans of Nainital and Ramnagar Forest Division.

4.2.1. Sal Forest

On the basis of lithological formation and the moisture content the Sal forests are divided into four types:

(a) *Dry Dun Sal:* This type occurs in Dechauri and Kota forest ranges in substantial portions of Mushabangar, Patkot, Bhalaun and Bhandar Pani. The soil is usually deep alluvial, clayey loam porous and well drained. It is very well suited to the growth of the Sal in these parts where Sal forms 90% of total area but Dhauri and Haldu are the principal associates. Others are Rohini, Khoda and Dudi, Gauj, Malu and Maljhan are common climbers.

(b) *Moist Dun Sal:* It is occupied by dun floor in Dechauri Kota, Mushabanger, Patkot and Bhalaun. The soil is usually deep alluvial, clayey loam porous and well drained in this region. The principal associates are Haldu and Dhauri. Jamun occurs along the water courses in dense. It is associated with Rohini, Khoda and Dudhi. The undergrowth is absent and duri and Gauj the most common climbers while Maula and Malijhan also occur to some excellent.

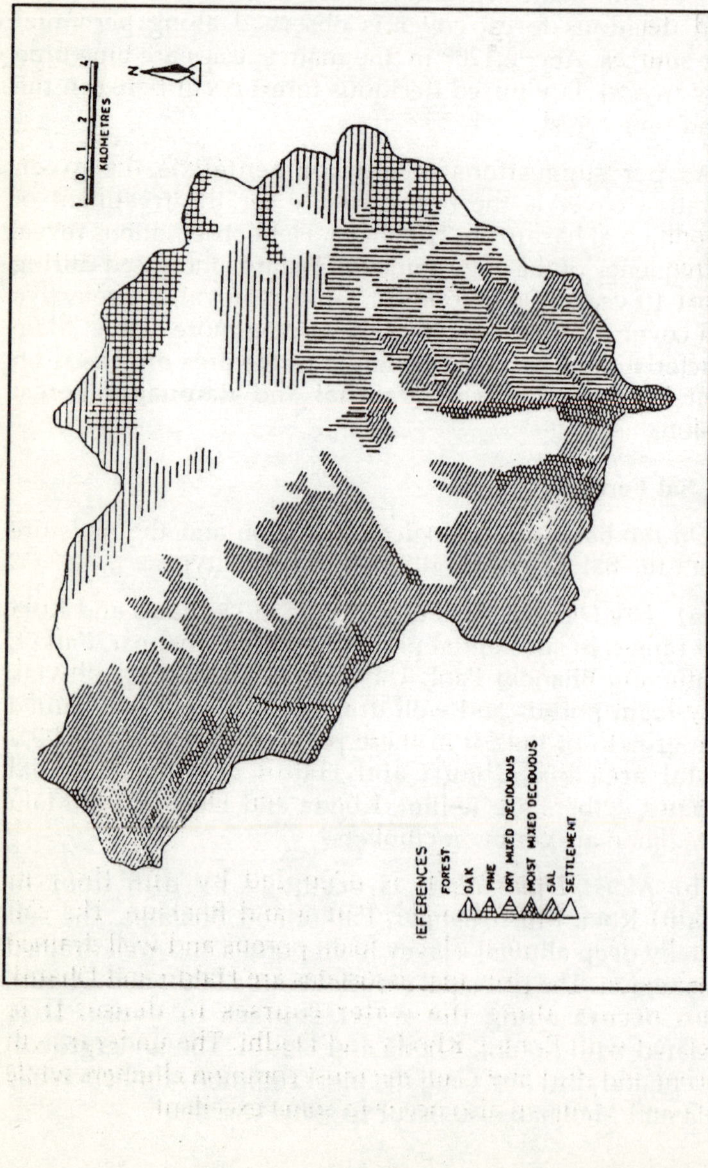

Fig. 4.2 Kotadun: Forest Types

(c) *Moist Siwalik Sal*: It occurs in foot hill areas of Fatehpur, parts of Selsia and Kotabagh. The subtype exists on greywack rocks with light soil. The best growth is found in Champa-Gad area. While the quality of Baluti sal is poor. Except Sal, Bahera, Jamun, Haldu, Titmira, Dhauri, Harar etc. also occur in lower hills near dun floor. Lower Siwalik includes Sandon, Rohini, Amaltas, Chiller and occasionally Aonla however in damper sides.

(d) *Dry Siwalik Sal:* This type is generally over an altitude varying from about 400 to 800 m. The ground is generally cut by nalas. Soil is sandy. Sal occurs along the nalas and generally towards gently sloping areas where moisture is high.

4.2.2 Pine Forest

The principal forest species of Lesser Himalaya is pine and no other species occur in the top canopy. Towards lower elevation it is often replaced entirely by low miscellaneous scrubs and other broad leaves species. Towards the lower limits of its habitat it passes gradually into forests of Sal or of various miscellaneous species, and towards its upper limits it passes into Oak forests. Other factors affecting its distribution are burning and lopping. Burning encourages Chir to spread upward into the oak bush appears to have the opposite effect at its lower limits where in many places sal and other species are gradually spreading into the Chir forests.

4.2.3. Oak Forest

It occurs on about 2000 m. elevation geological formation has very little influence on the distribution of oak forest which builds up a good soil rich in humas. The ground is naturally well-drained and the Oak occupies all but really wet soil where decidous trees take place. Lopping is now extremely prevelent as a result of fast disappearance or deterioration over large oak forest areas. Although inflammable but in many localities these forest suffer a good deal from fires spreading up from the pine zone below. Other associates are Rhododendron and Ardoreum. There is much moss and lichen on the trees and in the monsoon a capious development of epiphytic ferns. In cooler aspects where it is associated with Tilonj, the latter tends to replace it. Banj is increasingly common eastwards.

4.2.4. Dry Mixed Decidous Forest

Such forest occur in steep hill areas where moisture conditions are not favourable for Sal. The greenary of these forests is changed to leafless during the dry season. The top canopy consists mainly of Bakuli or Sain which are mixed with Haldu, Khair, Sal, Jigan, Titmola, Tendu, Chamror, Aonla. Baib, Gonia and Kumeria are the common grasses. The regeneration of Khair and bakuli is very well.

4.2.5. Moist Mixed Decidous Forest

This type is found in those areas where moisture prevails and which remain fairly cool throughout the year. The soil is loamous and good water supply is available. The principal species are Seris, Tun, Jamun, Semal with occasional Sain and Bahera. The undergrowth is dense and consists of Bindu, Bhant and Gandhela.

4.2.6. Valley-Swamp Forest

This type generally occurs along valleys of perennial rivers the soil is loamous to sandy, the overwood is generally dense and mainly consists of Jamun, Gutel and Gular. Typical associates indicating the moist conditions are Kekra and Kaula.

Chapter 5

Soil Genesis and Erosion

5.1 SOIL GENESIS

Pedology is the science which studies the origin, properties distribution and best uses of soil. Scientific study of soil is necessary for good crop production and landuse planning. In a general way we may define soil as a loose material which supports vegetation. Dokuchaw (1883) defined the soil as the surface and adjoining horizons of parent material which have undergone more or less a natural change under the influence of water, air and various species of organ.

Thus soil is an autonomous natural body resulting from the interaction of the parent rock with environmental factors of climate topography, plant animal life. Pedologist realize the close geomorphological impact on soil formation. Digger *et al.* (1974) present a close impact of landform on laterrite tract in west Bengal. Similarly, Hilwig *et al.* (1971), Pofali *et al.* (1979) and Jalal (1976) have also found close impact of landscape on soil properties. De N. K. and Chanduri (1984) have also tried to see the impact of various physical factors.

5.2 IMPACT OF PHYSICAL LANDSCAPE ON SOIL

Soil is the product of underlying parent rocks. It is the result of forces of gravity and downward movement of fine rock particles. The movement may be caused either by erosional factors viz., water, frost and wind by biological ones viz. Construction of road or dams. Third factor of weathering process which is responsible of parent rocks, deforestation also creates some modification.

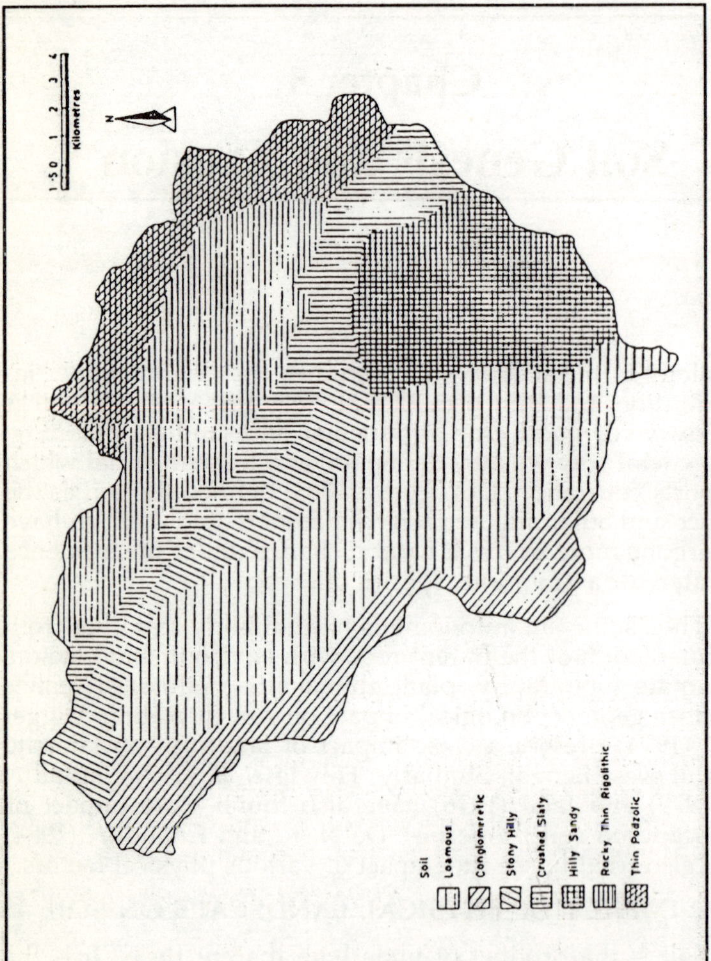

Fig. 5.1 Kotadun: Soils

In the absence of requisite information certain physical factors have been taken into consideration here in order to present the spatial distribution of soil. Although the mapping of the soil on the basis of geology, land-forms, climate drainage and vegetation is doubtful but these are important factors for determining the nature and evolution of soil. It is helpful in knowing the agricultural evolution resulting from the action of climate on a particular type of rock structure as the region of similar tectonic history and climate may have similar soil properties.

Based on lithology, altitude, landform, slope, temperature, vegetation and drainage density the soils of Kotadun are categorised into seven groups (Fig. 5.1). The grain size colour and grain types vary according to these factors. Affected soils are conglomeratic, crushed slaty and hilly sandy. Stony hilly soils are mainly affected by slope, and podzolic soil cover occurs due to presence of good forest cover. Regolith is found towards northern part where slope is moderate but altitude is high (Table 5.1).

A. Conglomeratic Sandy Soil

Presently it is covered with dense forests. Genegtic study indicates some hints towards the disintegration of recent deposits. Rainfall intensity as well as the porosity in this region is very high. Dense forest cover has dominant effect in the formation of this soil. In this region the average slope ranges from 7° to 21° and the elevation is 400 to 800 m. Rivers cast very deep gorges giving a clear indication of excessive soil erosion. Seasonal rivers are unable to form soil here because depositional process is not possible due to higher degree of slope.

B. Loamous Soil

Most of cultivated area of Kotadun lies under this soil. It is most fertile one. It extends east to west over the dun floor and is composed of sand pebbles. Pebbles are shorted out because of intensive cultivation. It is low land with an average slope of 7° to 14°. Drainage density and altitudinal

Table-5.1 Kotadun: Physical Characteristics of Soil

Soil Group	*Elevation (m)*	*Lithology*	*Landform*	*Slope*	*Temp (°C) min – max*	*Vegetation*	*Drainage density*	*Landuse*
Conglomeratic	400 – 800	Conglomerate	Low hill	7–21	12 – 33	Siwalik sal	Moderate to high	Fully forested
Loamous	500 – 700	Sand and pebble	Low and	4–14	13 – 34	Dun Sal	Low	Paddy, wheat, gram, lahi
Hilly Stony	900 – 1100	Slided mass	Low foot-hill	14–28	11 – 31	Sal Forest	Moderate to high	Mainly Wheat and Madua
Crushed	900 – 1100	Crushed slate	High hill	28–40	14 – 30	Dry mixed Decidous	Moderate to high	–do–
Hilly Sandy	500 – 1700	Greywacke & slate	High hill	14–40	7 – 24	Dry mixed decidous	Moderate to high	Mainly Ginger and wheat
Rocky Thin Regolithic	1100 – 2000	Quartzite	Mountainous part	28–40	8 – 22	Partially oak	Moderate to high	Ginger, paddy and wheat
Thin Podzolic	1800 – 2600	Dolomite	High mountainous part	28–42	8 – 21	Partially oak & Pine	Moderate to very high	Wheat and madua

Source: Field observations.

range are very low. As a result, the gravitational process is almost nil. The colour is brown to blackish. Organic contents are responsible for the blackish colour. Prior to settlement this area would have covered with dense sal forest. The effect of humas content is about 30 cm in depth while remaining depth is covered by gravely and sandy strata. Along river courses there are raukhers which are not cultivable because of excessive pebbles and cobbles. During rainy season these generally remain under floods for a period not exceeding 4 to 8 hours a day depending upon the intensity of rainfall. Except in rainy season the area is not drained by perennial stream except Kaligad and Baur. Thus soil forming process depends totally on human activities. This soil group contains 106.22 km^2 or 29 per cent of total area. It is mostly under cultivated area. The villages of Patkot, Chopra, Leti, Kotabagh, Nauda, Keshabanger and Patliya lie in it.

C. Stony Hilly Soil

These characteristics are generally observed near the northern boundary of the dun floor. This part is fully covered by the debris of old landslides which would have occurred due to succession of Siwalik ranges. The slided mass has covered the Lower Siwalik deposits so larger grains are of quartzite and smaller ones are of greywackes. Because of steep slopes and higher altitudinal ranges there is poor development of vegetation cover consequently soil is denied of organic content. The colour is light grey to brown. It is generally used for cultivation of Madua and Wheat. This category includes 29.72 km^2 or 8 per cent of total area in the villages of Boharakot, Devipura and Pandegaon.

D. Crushed Slaty Soil

This unit occurs mainly along the Main Boundary Fault, the contact zone of Lesser Himalaya and Siwalik ranges. Parent rocks of this soil belong to slates. Because of high compressional forces the whole formation of the area is crushed in fragmented fine chips. The soil erosional process is dominent in this zone of high slope and altitudinal range but mainly due to occurrences of a high sensitive zone of the M. B. F. If we observe towards genetic formation of soil, slope has no effective impact on its formation. The uppermost surface goes on washing out by and large but the parent rock being of crushed material some characteristics go on developing. This is the most hazardous zone of Kotadun where people suffer from mass wasting process of landsliding, slumping, soil creeping, mud flow etc. The grain sizes of the soil vary from 1.5 cm to 3 mm. Vegetation cover is in 15% of the area which is further decreasing because of increasing demands of fuel and fodder. The frequency of mass wasting process remains highest during the rains on account of increasing pressure of human as well as animal population and consequently rapid deforestation process have been responsible for increasing the mass wasting process during the last two-three decades. The soil making process thus is controlled mainly by the geological formation. This category soil spreads over 37.97 km^2 or 11% of total area. Main villages under this formation includes Goriadeo, Semalkhet and Adhaura.

E. Hill Sandy Soil

Parent rock of this soil falls under Siwalik formation consisting mainly greywackes. Altitudinal range varies from 500 to 1700 m. having 14°-40° slope. This zone is covered with the material of disintegrated loose greywackes of sandy nature and brown to greenish colour. Drainage density is very high but these drains are mostly seasonal. Porocity of the soil is high while moisture content is very low. Mineral contents containing soil fertility are good but texture is so coarse. This soil formation covers an area of 48.22 km^2 or 13 per cent of total area in the villages of Raukher, Nalni and Khamari.

F. Podzolic Soil

This zone is partially covered with oak forests under which decreased Lichen and Moss give some indication of podzolic natural of soil. Some where there are only bare rocks with high altitudinal range and slope. High moisture contents and thick litter of forests provide high humas content and as a result, soil colour is blackish to light grey. Main localities are along the uppermost part of Sigarhi and Ghugu Khan villages. The area of this group is 36.46 km^2 or 10 per cent of total area.

G. Rocky Thin Regolithic Soil

This soil is spread over those areas where forest cover is very low while altitudinal range is very high. Thus, the deposition of soil particles is not possible and only bare rocks are visible there. The deceased particles stop in the areas where slope is low. The colour is brownish and the grain size is coarse. Main villages in this group are Kalsi, Amota, Bansi and Salva. It covers 62.97 km^2 or 18 per cent of total area.

Table 5.2 Kotadun: Physical Properties of Soil

Soil Type	*Grain Size (mm)*	*Soil Colour*	*Grain*
Conglomeratic sandy	2– .2	Brown	Crumbly
Loamous	.02– .2	Dark Brown	Grainular
Stony Hilly	.5–1.5	Whitishgrey	Blocky
Crushed Slaty	1.5–3	Greenishgrey	Platty
Hilly Sandy	.3– 2	Brown	Grainular
Rocky thin Regolith	.5– 2	Whitishgrey	Grainular
Thin Podzolic	.02–1	Dark grey	Grainular

As regards the quality based on texture, the best soil is one that increases the areation and the water holding capacity and facilitates the activities of micro-organisms. Loose friable soils are easy to cultivate than heavy compact soils. In the study area loamous soil seems to be a good quality soil.

This study reveals that geology plays a significant role particularly on conglomeratic soil the basel part of which belongs to Upper Siwalik and crushed slaty soil along M. B. T. Regolith along moderate slope groups, stony hilly soil along steep slope and lomous soil along level area indicate the slope impact. Forest cover has also a significant impact on the podzolic soil formation along the high altitude. Thus soil genesis study includes a tremendous geomorphological attributes.

5.3. CHEMICAL ANALYSIS OF THE SOIL

Throughout history soil has been used by man for agricultural and structural purposes. Soil problems, as they occur in history, relate to changing quality or quantity. Agriculture may lead to changing soil quality by putting an excessive demand on plant nutrients. As a result the soil declines in its productivity. In order to maintain productivity, cultivator add chemical nutrients (fertilizers) that may adversely affect the plant and animal of the soil community. Irrigation, under conditions of poor drainage has led to the concentration of salts in the soil, making it toxic for most plants. By the way of exposing and cultivating the soil in turns, the quality of the soil goes on changing. The effects of the wind and water on exposed soils have led to serious problems of soil erosion. Changes in ground cover and landuse may lead to changes in the soil moisture regime and, therefore, the strength of the soil. Some clays, shocked or disturbed in the presence of moisture, well liquefy. The phenomenon has led to disastrous mudflows and mudslides.

Material and Methods

Soil samples from selected fields of village Patkot were collected in polythene bags and brought to the laboratory. The samples were obtained from the top of the soil surface and bottom samples were collected from 15 cm soil depth. Soil reaction was analysed by a degital PH meter, using 1:2 proportion of soil and water. Available phosphorus was determined by Kolt colorimeter, while Calcium, Magnesium and Magnese were analysed by Atomic Absorption Spteo Photometer in the exchangeable available form. The instrumental condition for analysing Ca, Mg and Mn was as below:

Calcium:	Principal line	–	422.67 nm
	Slit settings	–	0.4 m,
	Flames-Air	–	5.0 litre/minute
	Acelytene gas	–	1.0 litre/minute
	Lamp current	–	Maxim-10 MD
	(allow 80% current)		
Magnesium:	Principal line	–	258.21 nm
	Slit settings	–	0.4 nm
	Flames-Air	–	5.00 litre/minute
	Acelytene gas	–	1.1 litre/minute
	Lamp current	–	Maximum-4 MA (allow 80% current)
Magnese:	Principal line	–	279.48 nm
	Slit settings	–	0.2 nm
	Flame-Air	–	5.00 litre/minute
	Acelytene gas	–	1.00 litre/minute
	Lamp current	–	Maximum-12 MA (allow 80% current)

From a chemical view point soils are composed of a large number of constituents ranging from simple salts to highly complex inorganic and organic compounds. Many of the compounds occurring in soils are only poorly understood. Chemical processes continue unceasingly, with the result that soil composition is constantly changing (Lutz and Chandler, 1946). It is a fundamental fact that plants require a source of chemical composition if they are to develop normally. The major nutrient elements are nitrogen, phosphorus, potassium, calcium, magnesium, sulfur and iron it is now recognised that minute amounts of other element, such a boron manganese, zinc and copper are essential for many plants.

Present study deals with the analysis of calcium, Magnesium, Maganese, Phosphorus and Hydrogen-iron concentration of the soil of Patkot village in Kotadun (Table 5.1).

Calcium and Magnesium

Ca and Mg are very important elements in the life of soils. Their effects upon the vital activity of plants are both direct and indirect. Calcium improves the structure, physical and physico chemical properties of soils and encourages greatly the biological processes in soils, especially .the accumulation of saturated humus substance, while magnesium is a component of chlorophyll.

In surface soils the content of Ca usually varies from less than 0.1 to more than 5.0 per cent, while the Mg content in the upper soil layers varies from less than 0.1 to more than 2.5%. Presented data of exhangeable available Ca and Mg shows that the Ca content varied from 46.1 ppm to 52.2 ppm. The average concentration is higher in the bottom soil and the Mg content ranges between 32.4 ppm and 33.8 ppm. The average concentration is higher in the top soil.

Maganese (Mn)

In the top and bottom soils of various fields, the concentration of exchangeable available Mn ranges between 2.8 ppm and 42.6 ppm. Table 5.3 shows that the average concentration of exchangeable Mn increases rapidly with increasing soil depth.

Phosphorus (P)

Phosphorus in soil is represented by its organic (asptytin, phospholipeds and nucleoproteins) and mineral compounds (as apatite, a phosphate of calcium containing a small amount of fluorine of chlorine I. Mineral compounds of phosphorus in soils can be in the bonded, exchange and insoluble form. Because of the great variations in chemical composition, average values of phosphorus content are of little significance. However, as a very rough example, it may be stated that the amount in the surface soil generally varies from less than 0.01 to perhaps 0.20 per cent.

The available P concentration varies in the range of 7.00 kg/ha and 49.00 kg/ha. In the top soils, it increases markedly in some fields, while in bottom ones, it is not well marked from field to field.

Table 5.3 Patkot: Chemical Analysis of the Soil

Sl. No.	*Plot No.*	*Sample Top/Bottom*	*Ca (PPM)*	*Mg (PPM)*	*Mn (PPM)*	*P (Lg/ha)*	*PH*
1.	1	(T)	51.1	32.9	5.7	25	7.81
2.		(B)	47.9	32.7	2.8	15	7.68
3.	33	(T)	51.0	33.0	14.4	15	7.76
4.		(B)	50.3	32.6	4.2	15	7.75
5.	344	(T)	47.3	32.7	4.3	15	7.60
6.		(B)	47.1	32.4	5.3	10	7.30
7.	503	(T)	46.1	32.5	9.6	9	7.84
8.		(B)	49.9	33.4	42.6	8	7.75
9.	1321	(T)	47.1	32.8	8.3	12	7.75
10.		(B)	47.8	33.2	26.1	7	7.30
11.	1970	(T)	46.6	32.9	5.8	10	7.83
12.		(B)	50.3	33.8	8.5	20	7.75
13.	2396	(T)	49.3	33.1	5.4	49	7.90
14.		(B)	52.2	33.2	12.6	10	7.76
15.	2646	(T)	51.8	33.1	6.0	9	7.30
16.		(B)	51.6	33.4	6.5	10	7.25

Analysed in Soil laboratory, Nainital.

Hydrogen-ion Concentration (pH)

The most common method for expressing the relative acidity or alkalinity of a soil is by the pH factors, which is based on the relative concentration of active hydrogen-ions (H^+). With increasing concentration, acidity increases and pH decrease. Most soils range between 4 and 9 pH value. The pH values below 7 indicate acidic conditions, whereas from 9 to above indicate alkaline conditions. Nutural conditions are expressed by a pH value of 7 and it is the range between 5.5 and 7.5 that is most suitable for plants. In Patkot the pH of soil varies from 7.3 to 7.9, which is higher in top soils and lower in bottom ones. The soils in Patkot village are nutural or slightly alkanine.

5.4 ASSESMENT AND DISTRIBUTION OF EROSION INTENSITY

Soil erosion is a great environmental problem. Although man himself is not physically removing the soil but his mismanagement is provoking such problems. Sheet erosion and side erosion of river banks is prevalent in dun floor but landsliding and slumping activities are occurring towards higher areas. The hazardous soil erosion is observed along M.B.T.

The term soil erosion is defined as the erosion of soil by various natural geomorphological process. Except in certain circumstances man himself is not removing the soil but his mismanagement of the land allows natural processes to act with added vigour. Soil erosion therefore is accelerated by his increased activities. However the impact of man is a crucial factors affecting soil loss. Soil erosion is thus very active in many places particularly areas of (a) mismanagement of both arable and grazing lands, (b) deforestation on mountain slopes, and (c) execcessive construction of roads. Soil erosion is largely determined by the factors relating to climate, topography, soil characteristics, vegetation cover and landuse practices.

Present work is based on surface height variation these with the assumption that higher is the height variation higher is the erosion intensity. Dove Nir (1957) formulated the method of calculating the erosion intensity as follows:

$$D.I. = \frac{Rr}{Aa}$$

Where D. I. is dissection index Rr is relative relief and Aa is absolute relief.

The area is divided in one Km^2 grids and absolute and relative height is placed therein. Values were changed according to the formula of erosion intensity and arranged

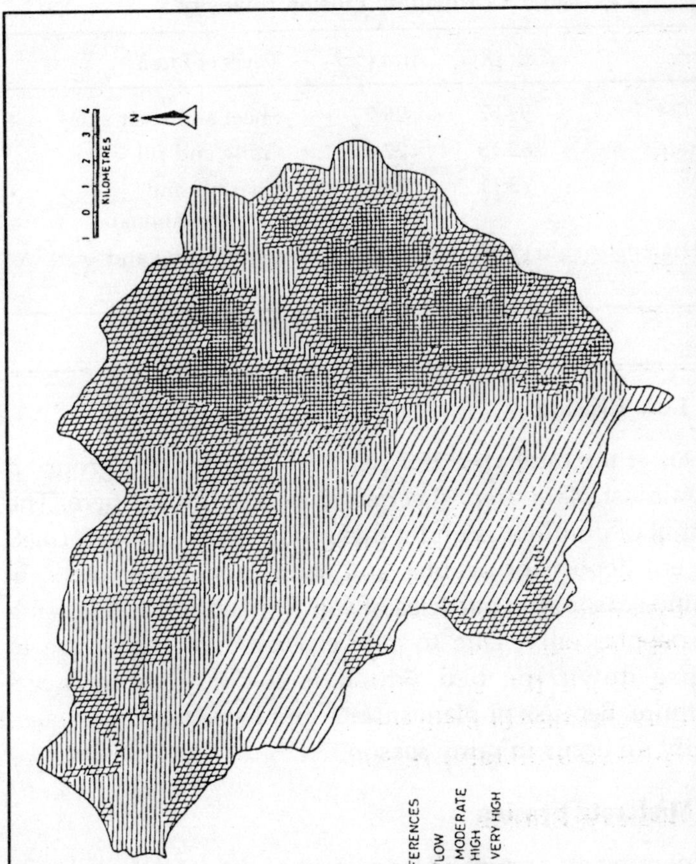

Fig. 5.2 Kotadun: Erosion Intensity

in four category isopleths. The area of each category is computed with the help of planimeter. Preparing this map erosional processes are observed on the field. The intensity is arranged in four categories (Table 5.4).

Table 5.4 Kotadun: Erosion Intensity

Category	*Area (Km)^2*	*Area (%)*	*Types of Erosion*
Low	92.57	25.7	Sheet and river side
Moderate	82.15	22.8	Gully and rill
High	73.11	20.2	Channel and gravitational
Very High	112.95	31.3	Landsliding and slumping
Total	360.78	100.00	

5.4.1. Low Erosion

Out of the total area 25.7 per cent is under this group. It being almost plain area, sheet erosion is dominent here. The litho-unit of dunfloor is under this group. Dun floor is formed by recent deposit of pebbles and sand carried by rivers. In the rainy season fine particles of soil flow with water leaving there pebbles only. Due to side erosion river sides use to collapse down the bed which is a curse to terraced agriculture. Because of plain surface and loose formation such calamity do occur in rainy season.

5.4.2. Moderate Erosion

Covering an area of 22.8 per cent of the total, this group includes Upper and Lower Siwalik ranges the lithology and structure of which is similar to dunfloor the compactness of rocks and the altitudinal variation of the former is higher than the latter. Both the gully and rill erosions are observed here causing frequent earth flow, mud flow and slumping of loose material.

5.4.3. High Erosion

Comprising 20.2 per cent of total area, the lithology of this group is compact but the higher slope bears maximum probability of soil erosion. Little quantity of water is able to do acceleration, of soil erosion. This is the cause of out washing of bare mountain expresses during rainy season.

5.4.4. Very High Erosion

Along the MBT this category is comprised of 31.3 Per cent of the area. Although the relative height of the area is not high but the existence of M. B. T. creates disastrous problems of soil erosion. The masswasting process during the S. W. monsoon period reaches to the climax. More than 90% of the hazards of Kotadun take place in this zone.

According to Vink (1981) soil erosion is of two categories: (i) man induced and (ii) natural or normal.

5.5 PREVENTIONS

1. The ideal objective of soil conservation may be obtained by conserving existing forests and planting all bare slopes with indegenious plant species.
2. Optimal landuse model should be implemented into action for reducing the effect of accelerated soil erosion.
3. Good turf of grasses should be developed for protecting the surface the interception of raindrops. It prevents the rainsflash erosion.
4. Contour ploughing and contour planting reduces the erosion intensity on mountain slopes.
5. Deep ploughing is important for increasing the infiltration of rain water. It reduces the surface run off and the erosion intensity.
6. High soil erosion is observed on fallow and wastelands. Recent reclamation techniques for these lands may reduce a reasonable effect of soil erosion.

Chapter 6
Specific Study of Landslides

Along Main Boundary Fault (M.B.F.) which is the contect zone between Siwalik and Lesser Himalaya, 17 landslides categorized into five groups—Khairar, Gadela, Sanarha, Dabka and Baur are studied in Kotadun. These are again divided into 7 primary and 10 secondary landslides. Steep slope, high relative relief, weak geologic structure, natural springs, unscientific construction and deforestation are found main causative factors. Affected afforestation programme masonry work, ashphalt mulch technique and contour agriculture are suggested as the preventive measures.

The term landslide is used for the processes of mass movement in loose material. Due to high relative relief acceleration provides high potential energy to little amount of water which falls in the area. Other associates which provoke the landslides are tectonic unstability of geological traits and peculiar climatic and anthropogenic process.

Such problem is reported by Ball (1878), Theobold (1880), Middlemiss (1890), Oldham (1893), Holland (1896), Banerjee (1955), Krishnaswami *et al.* (1964), Fookes and Wilson (1966), Nigam (1968), Dave (1970), Hukku and Narula (1975), Valdiya (1985), Bhandari (1985), Haigh (1979, 1984) and Jalal (1985) in Indian Himalaya.

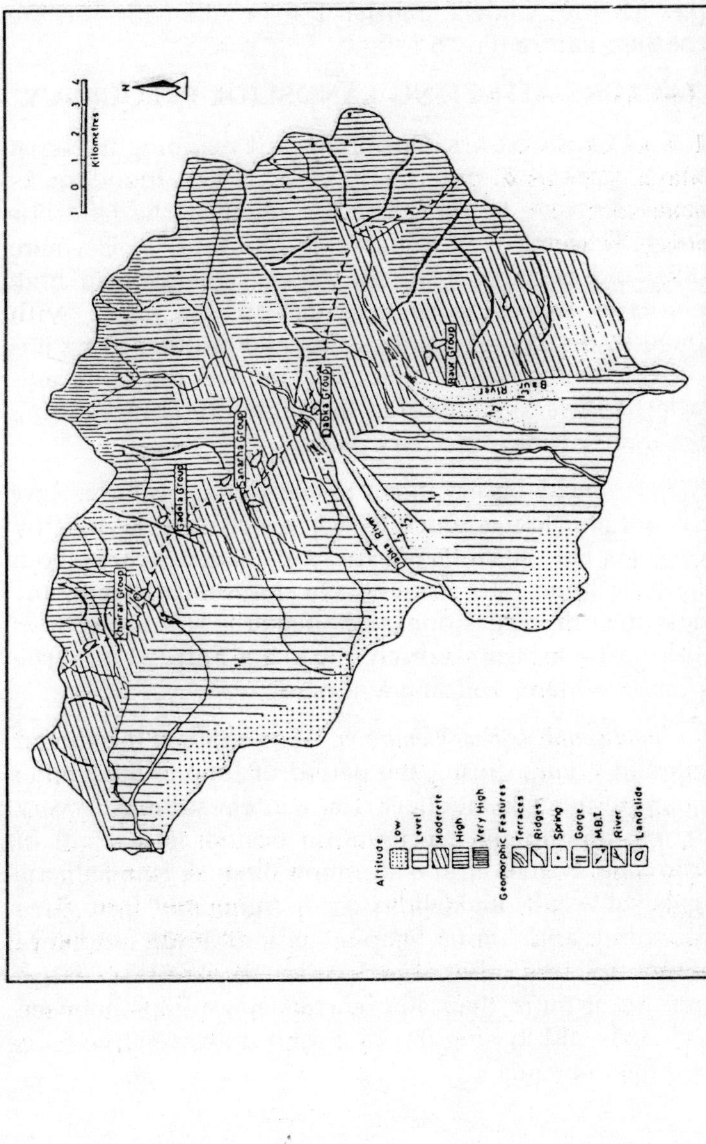

Fig. 6.1 Kotadun: Environmental Hazards

Here an attempt is made to study landslides based on detailed field investigation and morphological measurements with measuring tape, aboney level and brunten compass. After geomorphological mapping the small and large landslides observed in Kotadun are categorized into five groups—Khairar, Gadela, Sanrha, Dabka and Baur—on the basis of their nature (Fig. 6.1).

6.1 FACTORS AFFECTING LANDSLIDE FREQUENCY

1. *Rock Characteristics:* During hazard mapping in Nepal Himalaya Markus Zimmermann *et al.* (1986) found rocks characteristics very important factor affecting the landslide frequency. It seems very remarkable along M. B. F. where thrusting and faulting rocks are highly crushed and high slope creates landslide process easily associating with precipited water. Because of thrusting and faulting along this zone crushed material of slate, fragmented quartzite and grewacke is observed there as the high pressure bedding plane is changed into shistosity plane.

2. *Slope:* Field observations reveal that possible failure plane is not affected as much by weak structure as done by surface slope. For the landslide study we have taken the both side natural slope into account. On mountain terrain, the analysis of both side slope and unstable slope provides valuable information which may be helpful for the construction of dam, building and road.

3. *Climatic and Vegetal Condition:* Above 85% of the annual precipitation occurs during the period of June to September on account of monsoonal effect. The winters are mostly very dry. A few amount of precipitation occurs as a result of westerly wind system. In this duration Binayak Kunjakharak zone gets snow fall. Landslides occur frequently from June to September and about 90 per cent of total landslide frequency exclusively occurs due to rainfall. High temperature of more than 40°c creates space gaps between two rock beds and loosing the joint set but low relative relief decrease the movement.

The vegetal cover is about 10 per cent near landslide prone area. Although, the vegetal cover can not check the landslides completely but the landslide frequency may be decreased upto 40% through afforestation.

4. *Impact of water on the Landslide:* Hock and Bray (1977) observed that a little amount of water can provoke the slope failure process. High hill slopes tend to high frequency of landslides although the water impact is also considerable. Generally, water follows the weak structure plane, reduces the friction and becomes a lubricant material. The slope may thus, be destablized creating catastrophies. Natural springs along the weak and active plane at Khairar and Gadela are functioning as a lubricant. The landslides did not occur at Semalkhet before the construction of the canal but a patch of land used as kitchen garden is sliding downward continuously. Natural springs at Sanarha Talla are also functioning as lubricants, at this place the landslide in 1986 is triggered due to bedding plane failure.

6.2 CASE STUDY OF LANDSLIDES

(a) *The Khairar Group:* The landslides of this group six in number are observed near Khairar and Goriadev villages. The nature of these landslides is almost similar. These make a cluster around Khairar badly affecting the social economy of the residents. On account of occurence of Main Boundary Fault, the rocks are composed of crushed material. The vegetal cover is reduced badly by the natives for meeting their fuel, fodder and timber requirements. A few trees of Mango (*Magifera indica*), Haldu (*Adina cordifolia*) and Kusum (*Schilchera trijuga*) stand sparcely.

The main Khairar landslide lies between Ria and Kalmetia streams at an elevation of 1100 m. above m.s.l. The slope of the locality averages to 60°. The crown of the slide is located at a height of about 320 m from the base. The affected area extends longitudinaly for about 200 m. and transversely for about 150 m. The surface is characterized

Table-6.1 Kotadun: Primary Landslides Along Main Boundary Fault

Sl. No.	Variables	Khairar Group	Gadela Group 1	Gadela Group 2	Dabka Group 1	Dabka Group 2	Sanarha Group 1	Sanarha Group 2
1.	Lithology	Quartzite	Dark Brown Slate	Dark Grey Slate	Slate & Quart-zite	Quartzite	Greywacke	Quartzite
2.	Width (m)	30	22	16	70	25	30	30
3.	Length (m)	30	22	25	40	35	40	100
4.	Height (m)	23	18	19	35	30	32	70
5.	Slope (°)	55	50	50	45	50	50	65
6.	Area (m^2)	900	484	400	2800	875	1200	3000
7.	Volume (m^3)	91	1132	632	22295	2437	3744	8190
8.	Nature of slope	Concave	Concave	Concave	Concave	Concave	Concave	Concave
9.	Length: Toe to Crown	60	60	30	100	200	50	200
10.	Slope: Toe to Crown	60	65	55	50	65	55	60
11.	Natural slope (°)	45	60	75	70	50	60	45
12.	Dip & Direction of Beds (°)	19/20	50/20	20/30	77/180	57/160	86/290	54/35
13.	Dip & Direction of Joints (°)	–	75/70	82/120	–	–	–	–
14.	Height: Toe to Crown	50	50	25	65	120	35	160
15.	Orientation (°)	320	290	215	130	300	110	115

Source : Personal fieldd study.

Table- 6.2 Kotadun: Secondary Landslides Along Main Boundary Fault

Sl. No	Places / Variables	Khairar Group 1	Khairar Group 2	Khairar Group 3	Khairar Group 4	Gadela Group 1	Gadela Group 2	Baur Group 1	Baur Group 2	Sanarha Group 1	Sanarha Group 2
1.	Lithology	Crushed Material	Crushed Slate	Crushed Slate	Dark Slate	Crushed Slate	Dark Grey Slate	Greywacke Slate	Greywacke	Crushed Rocks	Crushed Rocks
2.	Width (m)	35	45	25	20	28	30	90	40	40	20
3.	Length (m)	120	40	40	10	40	40	30	50	60	40
4.	Height (m)	80	30	30	8	30	33	20	40	45	35
5.	Slope (°)	50	55	45	45	45	55	45	35	55	55
6.	Area (m^2)	2800	1800	1000	200	1120	1200	2700	2000	2400	800
7.	Volume (m^3)	12740	7879	2437	416	3057	3861	21060	8320	9360	1820
8.	Nature of slope	Convex	Convex	Convex	Concave	Concave	Concave	Concave	Convex	Convex	Convex
9.	Length: Toe to crown (m)	240	90	70	40	125	140	150	120	200	50
10.	Slope: Toe to crown (°)	60	60	55	50	60	65	65	60	50	65
11.	Natural slope (°)	60	55	50	55	60	70	45	45	60	50
12.	Dip and direction of Beds (°)	..	..	..	..	..	..	..	..	..	..
13.	Dip & Direction of Joints (°)	..	..	..	..	..	60/40	..	..	..	..
14.	Height: Toe to crown (m)	200	70	60	40	100	90	100	90	200	30
15.	Orientation (°)	40	310	320	100	30	130	160	100	80	350

Source: Personal fields study.

with a number of terraces which indicate the unstability of the area. The head scarp is fairly steep and free from vegetative cover. Towards the toe of the slide there are some springs and seepages which indicate that the impervious slip surface retaining the surface water hits the ground surface near the toe of the slide.

(b) *The Gadela Group:* The mighty nature in the night of 8-9 September became curse for six persons who got permanent peace in the arm of nature and burried under three meter thick debris mass of landslide. Ten days before the accident there was a clear indication of landslide by 6 inch wide crack at the crown part. On the 8th Sept., after 8 hours continuous rainfall, it suddenly became a slip plane by the night other landslides also occurred towards its left side. In this group four landslides were studied deeply.

Near about 118 m from the crown river Kaligad flows although there is no direct impact. These four landslides have different orientation. Except Gadela first (Table 6.1 & 6.2), all these are primary landslides.

(c) *The Sanarha Group:* This group also lies within the very active zone of M. B. T.. Steep slope, high relative relief, weak structure, natural springs, unscientific constructions (houses, terraces, canals and paths) and deforestation are the main causes for these landslides. The condition of these slides is becoming bad to worst. This group includes the landslides in Semalkhet village and those along Semalkhet Gad at three locations.

Lithology of this group is also very weak. Greywacke and Slate in crushed form is present everywhere. The bedding dip amount, strike line and dip direction are 83/15-1951—290 respectively.

The vegetation species are Jamun (*Syzygium cumini*) Lentana, Banana, Tun Dhaul, Tilpher etc. The slope is southern and sunny.

(d) *The Dabka Group:* This group belongs to the upper course of Dabka river causing enormous side erosion. The study of these landslides is essential in order to assess some remedial measures for proper utilization and management of water resources.

This group includes three active landslides along the Dabka. The crown of the slides is located at a height of about 100 m. above the river bed. The affected area extends longitudinally for 75 m. and transversely for about 100 m. The head scarp is steeply sloping and without vegetative cover. Active slides are forming a convex slope of 60° creating dangerous condition of the crown part.

(e) *The Baur Group:* Landslides are so frequent here that hardly any year passes without catastrophies. Along the steep sides of Baur river the softer rockmass and the weathered rock mantle slide shown under the effect of gravity. Such landslides are triggered due to percolation of rain water lubricating the weaker planes. The tie up of geologic structures with the intensity of climorphogenetics have given favourable condition for the origin of peculiar mass wasting process.

6.3. ANTHROPOGENETIC PROCESS

Although there is a natural adaptation by the natives of the affected area after a long experience but it is unscientific one being responsible for catastrophies. Deep observations reveal that canal, building unpaved path constructions and unscientific agricultural operations are the main causative factors for landslides. Although it is a weak lithological zone yet the unscientific activities of the people are creating serious problems. At Sanarha Malla a steep slope is disturbed for construction of a path approaching Mahatgaon. The path is followed by a drain in rainy season as a result of which toe cutting and down ward movement of crown part occur frequently. Deforestation is also done in great amount for meeting the daily local requirements of fuel, fodder and timber. Other responsible factors are rough grazing and unscientific cultivation. Near village Sanarha the crown part is under intensive cultivation and the toe part is used as the place of fetching water from natural springs. Middle part of the village is used for settlement. Such adoptation is really very dangerous and is unwanted invitation to slope failure process.

Main cause of slope failure at Gadela is the construction of building and canal the concave slope is cut at right angle towards the toe. Associated factors including rainfall decreased the resisting capacity of the crown part. Path construction, grazing and deforestation were supporting factors.

All these anthropogentic processes may be summerized in three categories:

(i) Most important causative factor is the deforestation of land. According to the local people calamities related to slope failure have been occurring for a long time ago but the position deteriorated recently. It may be due to decreasing vegetative cover. In the same geocharacteristics towards the east dense forest cover has checked all these problems.

(ii) Settlement processes are also disturbing the structural formation, and weak structure with the effect of other associates creates landslide problems. People use to cut slopes at right angle for building while the loose material litho-units are unable to bear 90° angles.

(iii) Unscientific cultivation without vegetative cover is creating soil erosion.

6.4 THE CAUSES

1. All landslides are along the Main Boundary Fault at the contact zone of Lesser and Siwalik ranges. As a result active fault zone is working as a amphitheater.

2. Loose lithology associated with high gravity and rainfall decreases its resisting capacity.

3. Because of deforestation, water is percolating upto basal rocks and it is acting as lubricant for sliping down of the land mass.

4. Many rills and drains percolate their water on vegetationless secondary landslides and the pressure in the rainy season bursts down along original structure following surfacial slope.

5. Steep cutting of slope towards toe side for cultivation and house building, and over-grazing along sunny site.

6. High side and lateral erosion by rivers is also responsible for the expansion of slided area. The measurements taken during field investigation are divided into two groups of primary and secondary landslides (Table 6.1 & 6.2). Primary landslide denotes to the areas where the original slope is disturbed along a plane of hard rocks. Secondary landslide denotes the slope failures occurring along the already disturbed surfaces and the debris is again sliping gradually. Among the total 17 landslides studied in the area 7 belong to primary and 10 to secondary. With the help of corelation values it is tried to observe the relation of one variable with others.

High positive relation in primary landslide is seen between length and height, length and slope, height toe to crown and height of landslide cone, width and volume. Where negative set is poorly correlated.

Under the matrix of secondary landslide highly correlated variables are length and height, width and volume, height and length toe to crown, area and volume under positive set. Under negative set all variables are poorly correlated.

6.5 PREVENTIONS

In order to restore the deteriorated situation responsible for creating havok, afforestation is considered as the most preventive measure. A dense vegetative cover may help in preventing the gravity of rainfall and protects the surface from the velocity of runoff and the cutting action of water. It increases the porocity of soil providing good condition for vegetation growth.

It is very important to notice that the most hazardous position is created by human interference due to unscientific developmentary process particularly of house building and agricultural processes. Crown and toe balance is necessary. If only 4 per cent of the volume of the slide mass is displaced from the head to the toe, its stability increases by 10 per cent (Zaruba and Mencl 1969). So increasing pressure should not interfere towards crown part. Retaining walls should be prepared to protect changing angles of slope, particularly of house building process should be checked. Gadela landslide is the direct example of slope failure process because of disturbed slope angle. Slope may be reduced by the provision of terraces across the slope. Provision of terraces reduces the length of the slope and the amount as well as velocity run off is reduced and the erosion is checked. Single drain is developing as a well developed channel on account of combination of various rills. Rill development is based on uncovered surface. This problem may be checked by afforestation. Channel erosion should be controlled by well cemented diversions following topographic slope, for good outflow in rainy season. For commanding this problem qushing channels should be lined into pacca drains.

Along river banks, because of side erosion, the slumping process is frequently prevalent. For the effective preventive measurements checkdams may be constructed.

Canals should be well maintained and cemented because the percolation of water creates landslide becoming slipe plane lubricant passing through weak structural plane.

Agricultural fields should follow contours. Rocky area may be protected from the erosional problems by asphalt mulch technique. Bhandari and Gupta (1985) suggests spraying of the mulch at the rate of 0.15 to 0.2 gallon per hectare. The application of this method will create the moisture content as well as nutrients in the soil mental in order to conserve and promote vegetation and reduction of the susceptibility of slope erosion.

Chapter 7

Population, Settlement and Infrastructure

7.1 POPULATION

A geographical study of population concerns with various factors among which growth, distribution, sex composition, age group, literacy and occupational structure are the most important. The study of population is essential from the point of view of economic welfare of the society. There is usually a direct relationship between the population and intensity of prevailing landuse (Khan, 1965).

The landuse depends upon the objective of life and tradition of the people living in an area. Therefore the landuse gets modified on account of growing needs of population growth and literacy which are the instruments in bringing about agricultural change is an agarian society (Singh & Dhillon, 1984).

The distribution of population differs from one place to other. It is determined by various environmental factors. Physical conditions have both restrictive and permissive relation to human activities (Kayastha, 1964). Most of the population of Kotadun is concentrated in dunfloor owing to its favourable climatic conditions, availability of drinking and irrigational water and fertile soil while the mountainous part possessing adverse climatic conditions and steep inclines is sparsely populated.

The total population of Kotadun as per census of 1981 is 20, 657 which is entirely rural. The simple land man ratio of Kotadun is 57 persons/km^2 for the geographical area and 206 persons/km^2 for the revenue area.

Growth

As evident from table 7.1, the population of the region has a remarkable increase of 22.72 per cent in the decade of 1971-1981. Although it is comparatively lower than the national and district figures. The pattern of growth is not uniform all over the region. Within the span of ten years, the population growth in the lower units is higher than regional average while it is lower in the upper units.

Table 7.1 Population Growth (1971-1981)

Unit	*Population 1971*	*Population 1981*	*Variation*	*Percentage Variation*
U_1	594	691	97	16.33
U_2	365	415	50	13.70
U_3	773	832	59	7.63
U_4	1,279	1669	390	30.49
U_5	2,807	3423	616	21.95
U_6	638	875	237	37.15
U_7	3,137	3,647	510	16.26
U_8	1,548	1,877	329	21.5
U_9	186	233	47	25.27
U_{10}	5,505	6,995	1,490	27.07
Kotadun	16,832	20,657	3825	22.72

Source: District Census Hand Books, Nainital 1971 & 1981.

The areas registering a growth less than the average of the region, are contiguous to the dunfloor lying at the foothills. These areas are marked with less agricultural potentials owing to steep slopes, infertile soil and poor infrastructural development while the dun floor having greater infrastructural development and agricultural prosperity, have shown a mark of higher population growth (Table 7.1).

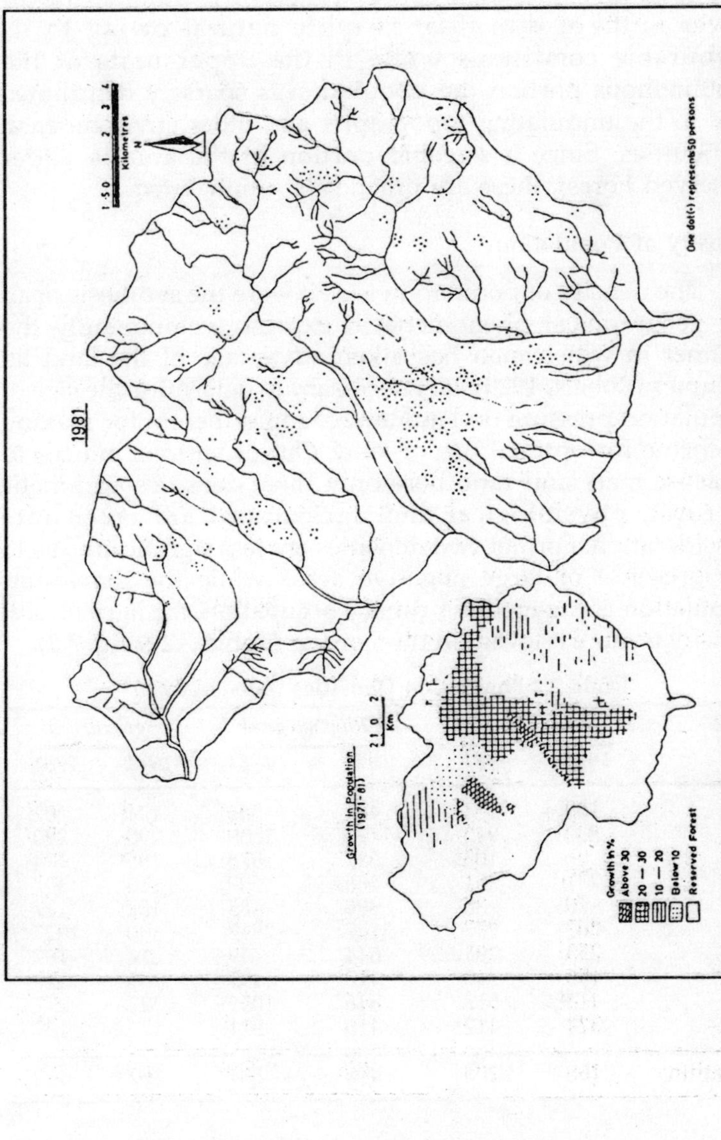

Fig. 7.1 Kotadun: Population Distribution and Growth

Distribution of Population

It is conspicuous from fig. 7.1 that the distribution of population is very uneven. The diversity of environmental factors in the region, has greatly influenced the distribution of population. The high concentration of population in the lower units of dun floor is quite natural owing to its favourable conditions while in the upper units of the mountainous portion the population is sparsely distributed due to the undulating topography and other environmental difficulties. Since a sizeable portion of the area is under Reserved Forest, these are practically uninhabited.

Density of Population

The density of population is as it were the synthesis of all the geographical phenomena: it expresses elonquently the manner in which man has taken advantage of the land he occupies (Robert, 1920). It is necessary to determine the size of population pressure on land and changes therein for making a scheme for optimal use of land. Out of various indices to measure man land ratio here only three densities-arithmetic or rural, physiological and agricultural are taken into consideration. Former two densities are less significant due to the presence of large negative areas while the maximum population is engaged in primary occupation, the agricultural density is more relevant in this context (Table-7.2 & fig. 7.2)

Table 7.2 Population Densities (*persons/km²*)

Unit	*Arithmetic*		*Physiological*		*Agricultural*	
	1971	*1981*	*1971*	*1981*	*1971*	*1981*
U_1	106	124	487	566	121	202
U_2	830	943	1177	1339	290	390
U_3	95	103	537	578	195	174
U_4	231	301	466	609	113	307
U_5	70	86	398	485	182	127
U_6	202	277	2127	2917	490	1027
U_7	253	294	644	749	214	187
U_8	198	240	410	497	183	128
U_9	413	517	845	1059	327	277
U_{10}	324	412	410	521	137	127
Kotadun	168	206	476	584	167	165

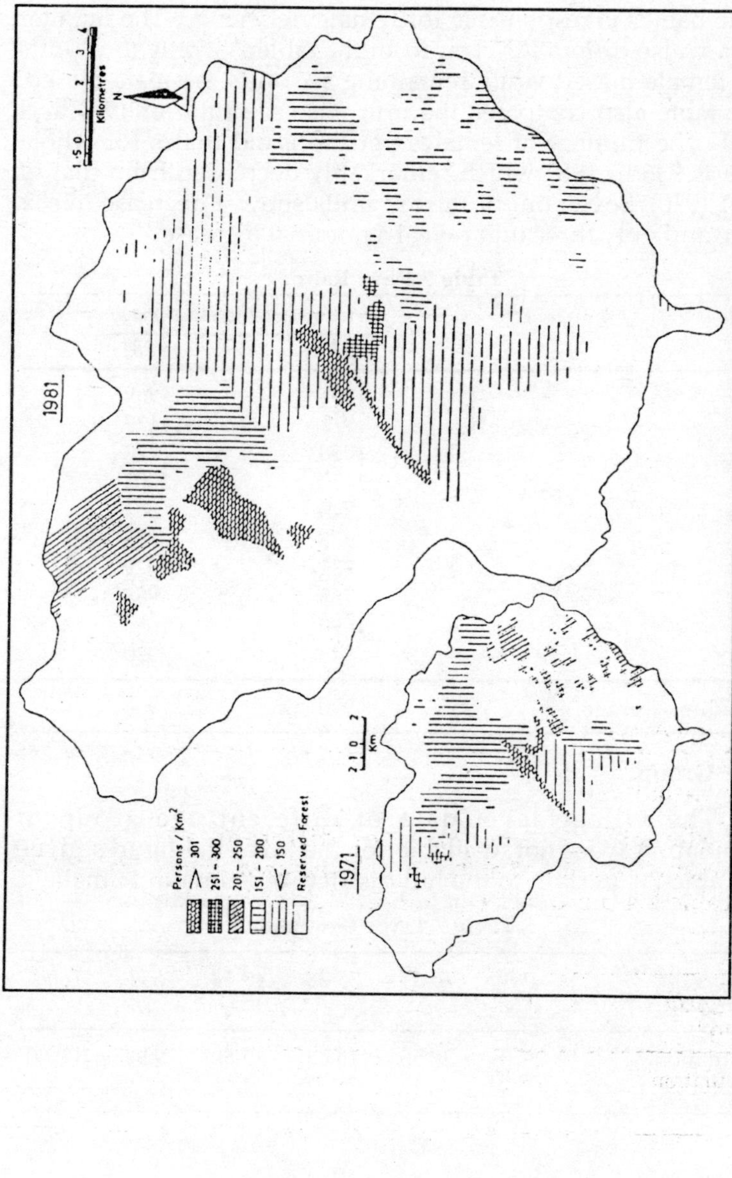

Fig. 7.2 Kotadun: Agriculture Density

Sex Composition

Table 7.3 shows that the male population is more than the female. The ratio varies from 882 to 1139 females per thousand males. Prevailing social tendency of favouring and welcoming male babies is responsible for female deficiency. The manner of care also differs. As clear from the table 7.3, only four units are female biased while remaining six units are male biased. The table also compares the unit wise sex ratio of 1971 and 1981. The number of females per thousand males for whole area is 936 in 1981 which remarkably decreased from that of 1971 (970). Seven out of the ten units show a decrease in sex ratio and only three unit reveal a nominal increase.

Table 7.3 Sex Ratio

Unit	*Females per thousand males*	
	(1971)	*1981)*
U_1	1055	969
U_2	921	1139
U_3	1050	972
U_4	1152	1003
U_5	940	882
U_6	975	953
U_7	932	1029
U_8	785	988
U_9	1090	1080
U_{10}	945	920
Kotadun Average	970	936

Age Group

The village level data of different age group of population were not available for the area, the figures given in Table-7.4 pertain to the hill districts of Kumaun Himalaya.

Table 7.4 Age Groups

Age group (Years)	*0-9*	*10-24*	*25-39*	*40-54*	*55+*	*Total*
Population (%)	29.18	28.58	18.82	13.80	9.92	100.00

Literacy

Level of literacy as a mark of social progress, is poor in the study area. It is 46.36 per cent in 1981. But literacy of 15.38 per cent of female population is lower than the literacy of 30.98 per cent of the total population in 1981. Unit-wise percentage of literacy is very uneven. The highest (69.95%) percentage of literacy in U_9 while U_3 stands the lowest (20.91%). The lower units possessing more amenities, are under high percentage of literacy while in the remote areas, the literacy is nominal. Unit-Wise literacy and percentage of increase during 1971-81 is shown in table 7.5.

Table 7.5 Literacy

Unit	*Percentage of literacy (1971)*	*Percentage of literacy (1981)*	*Percentage increase*
U_1	29.63	34.54	4.95
U_2	22.47	27.47	5.00
U_3	15.65	20.91	5.26
U_4	34.33	43.56	9.23
U_5	23.44	31.32	7.88
U_6	22.10	42.28	20.18
U_7	29.84	44.04	14.20
U_8	42.38	59.99	17.61
U_9	60.22	69.95	9.73
U_{10}	44.74	56.98	12.24
Kotadun	34.40	46.36	11.96

Occupational Structure

Among all the social attributes of population occupation is of paramount importance since it exerts vital influence on several personal, social and demographic characteristics (Gosal & Krishnan, 1965). In Kotadun where entire population lives in several areas, the occupation itself is often a reflection of a variety of cultural traits of the worker. Consequently, there are 38.71 per cent of the total population is worker and rest people are as non-worker.

As a rural residents, these people depend entirely upon agriculture. Out of the total working population 72.86 per cent are engaged as cultivators and agricultural workers. The house-hold industry, livestock, forestry, manufacturing, processing, servicing and repairing etc. engage only 12.47 per cent of the total workers. The remaining 14.67 per cent of the working population is of marginal workers. The simplicity of occupational pattern of rural populations is a reflection of their simple living and simple attainment.

7.2 SETTLEMENT

Settlement refers to people or colonies establishing themselves in a fixed position (James, 1934). Settlement in any area is the product of the cultural traits and time and provides the synthesis of dynamic pattern of inter-regional character of cultural landscape (Jordan, 1966). The nature and distribution of rural settlements of any area is largely influenced by the environmental conditions. The rural settlements is basically the residents of cultivators. The settlement pattern of an area is the result of reciprocity of spatio-temporal conditions.

The evolution of settlement is determined by the geographical as well as historical factors. After 1815 the region was under the British Division of the United Provinces of Agra and Oudh while before that it was under the rule of Chand Kings.

Right from the beginning, the emergence of settlement in the hill areas followed the agricultural practices. Chand (1986) described the settling process into four phase sequence of the settlement evolution in the Himalaya. Maximum inhabitants of the area came from entire and remote hill areas where there were no better prospectus for expansion of the cultivation. Since, dun area was densely covered by shrubs and bushes during that period, people cleared the unwanted weeds and shrubs, formed temporary residents and fields while twice or thrice time they suffered from cholora, plague etc. ephidemic and they were also ruled out from their properties by Gorkhas. However, they settled in nearer hill side and kept contact with the dun properties. Overall from a bird's eye view, the distribution of settlement is mainly concentrated in dun or along low lying areas because of their easy approaches, availability of water for drinking as well as irrigation, fertile soil and almost level land fit for intensive cultivation. With the increase in population, the land on the slopes and ridges in nearest approaches of these settlements, was also brought under cultivation and progressively came under settlement.

Consequently maximum settlements are found in dunfloor or lower units as compared to the slopes and ridges above those. As clear from the evidences of ruins of the houses, oldest settlement in Kotadun were near Devipura and Ranikota one km away from Kotabagh.

Distribution

The size of settlement largely depends upon the means of nourishment and economic conditions of the inhabitants. The evolutionary process of the settlement type varies from one place to another and it largely depends upon the interaction of several cultural and natural factors. In primitive stage people fulfilled their requirements, personally and they did not need to develop a business centre. All the 97 settlements are thus classified as rural settlements. Out of total, 96 villages are inhabited and one is uninhabited according to 1981 census. The size of settlements according to their population, ranges from 3 persons in Motibagh to 1,588 persons in Patlia. The cause of this variation is the unsystematic classification of the village by the Revenue Department since, only one hamlet has been considered as a village in certain cases whereas six-seven hamlets have been grouped into one village in many cases. The size of settlement is given in table 7.6 and depicted in figure 7.3.

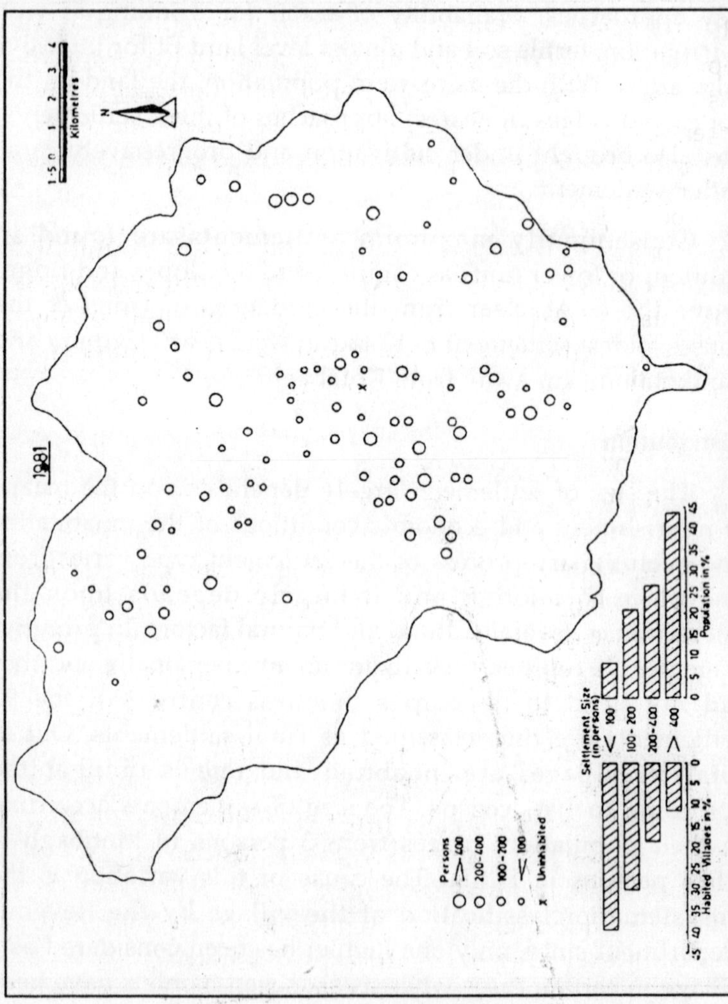

Fig. 7.3 Kotadun: Distribution and size of Settlements

Table 7.6 Kotadun: Size Distribution of Settlement

Sl. No.	*Size*	*No. of villages*	*% of total inhabited villages*	*Cumulative percentage*	*Total population*	*% of total population*	*Cumulative percentage*
1.	Above 400	11	11.47	11.47	9056	43.84	43.84
2.	200-400	18	18.74	30.21	5495	26.60	70.44
3.	100-200	29	30.21	60.42	4322	20.92	91.36
4.	Below 100	38	39.58	100.00	1784	8.64	100.00
Uninhabited		1	-	-	-	-	-
Total		97	100.00	-	20657	100.00	-

It is obvious from table 7.6 that 29.5 per cent or 6106 persons of the total population are concentrated in about 70 per cent or 67 villages while more than two third or 70.44 per cent of the rural population resides in only 30.21 per cent or 29 villages. It is very interesting to note that a large proportion of population is concentrated in few villages while a small proportion of that resides in many villages.

Types and Pattern

The degree of adherence and closeness of houses depends mainly upon the intra-house space relation, the arrangement and grouping of houses in a piece of land. In this way, the degree of adherence and closeness is very high adjoining dun-floor while the scattering of settlements is higher in the mountainous part.

The evolution of various type of settlements is an outcome of the cultural and physical environment of the region. Generally, dispersed and compact types of settlement are found in the region. Former type of settlement is found usually in northern mountainous part while the compact or agglomerated type of settlement is entirely found in lower units of dun floor.

The pattern of settlement means the special shape of the dwellings of a particular village which is mainly determined by the location of the village paths and the sequence and arrangement of the houses. According to topographical sheets as well as field observations, there are mainly linear, chackerboard and irregular pattern of settlements. Linear settlement pattern is observed along roads and canals in the dun floor where the natural springs are nil therefore, they are settled along canals. Checkerboard settlement pattern is also found in dun floor at crossing places of canals and roads. Irregular pattern of settlement is generally found in hilly portion where the topography has restricted the settlement pattern. The causes of such pattern of settlement are many fold in this area. Although the water supply, distribution of arable land and other infrastructural conditions are responsible for emerging such pattern of settlement in a small locality, which varies from one place to another. Caste system is also dominant factor which encourages the irregular pattern of settlements because, each caste lives in a group inherently forms one *'BAKHALI'* or *'DHARHA'* named after the caste. For instance, Nainwal Bakhali in Amgarhi, Pathak Bakhali in Bhalaun, Sahi Bakhali in Saur, Sati Bakhali in Patli while scheduled caste Bakhalies known a *Dumanrha* or Harizan Basti.

Generally the building material of the houses consists of locally available slate stone and timber in mountainous tract, while slate, iron rod, cement and timber are used in dun floor. Houses are single or double storied. The ground floor is generally used for stalling animals, although separate thatches and cattlesheds (*Gausala*) are made of locally available grasses, timber and red clay plaster. The settlement patterns are rapidly changing due the induction of new agricultural techniques and establishment of small scale industries or domestic industries.

7.3 LIVESTOCK

Along all cultural elements, intimate relationship between man and animals is very considerable, such type of subsistantial agriculture is observed in the study area as well as in the Himalaya. Livestock breeding plays significant role in agricultural economy due to the scarcity of arable land and less induction of new agricultural techniques. The number of animals depend upon the size of family and fodder supplies from the surroundings. Generally, every family has atleast two animals. Cattle are generally used for ploughing, Weeding carriers an drought animals, while meat and eggs are widely consumed. The total animal population grouped in four categories of bovine, ovine, equine and others is 25857 (Table-7.7).

Bovine

It includes cows, bullocks, buffaloes accounting 17,432 or 67.41 per cent of total animals while 40.60 per cent of the bovine are males and 59.4% are females, male bovine are generally used for ploughing. Keeping in view the manuring problem people keep larger number of bovine though this practice deteriorates the quality. The working efficiency of bullocks used in hills is very poor while is dun floor tractors replace the bullocks for ploughing, thressing etc. Even than each cultivator keeps one pair of bullock and one or two milch animals. People generally keep two buffaloes or cows for fulfilling their milk requirement.

Ovine

Including sheep and goats these animals account 4193 which constitute 16.21 per cent of total animal population. Goats and sheep raise the economy of the cultivators. These are generally used for obtaining meat and manure. Ovine are sacrificed to God and Goddes as per tradition of their worship. Besides it, these are also sacrificed to please the devil souls an order to rid over their bad effects. Goats and sheep are very harmful in the sense of environment because they are more selective feeders and to their easy access to trees for brauzing.

Table 7.7 Livestock, 1986

Unit	*Total* (No)	(%)	*Bovin* (%)	*Ovine* (%)	*Equine* (%)	*Poultry* (%)	*Others* (%)
U_1	884	3.42	3.37	5.56	1.70	1.69	0.86
U_2	525	0.03	2.51	1.03	0.71	1.51	0.43
U_3	1,338	5.17	5.01	7.12	5.45	3.66	2.59
U_4	2,098	8.11	10.12	4.84	3.21	3.70	1.51
U_5	4,855	18.77	17.85	24.76	25.00	15.46	9.48
U_6	1,251	4.84	3.79	5.75	0.36	10.71	11.96
U_7	5,507	21.31	19.41	32.03	30.54	13.68	15.09
U_8	2,566	9.92	11.21	7.18	9.55	3.66	13.36
U_9	295	1.14	1.08	1.07	1.34	1.51	1.51
U_{10}	6,540	25.29	25.65	10.66	22.14	44.42	43.21
Kotadun	25,859	100.00	100.00	100.00	100.00	100.00	100.00

Source: Block Head Quarters, Kotabagh, Bhimtal and Ramnagar.

Equine and Others

As indicated in table 7.7 these includes horses, mules, and pigs accounting a number of 2048 or 7.93 per cent of the total animals. Ponnies forming a predominant proportion, are used as luggage carriers to the interior villages.

Poultry

As shown in table 7.7 there are 8.45 per cent birds. Eggs and chickens are the main products of poultry, but the requirement of the region is mostly mitigated by imports.

The environmental condition is favourable for the development of animal husbandry but existing situation is considerable. Cow and buffalo are poor milkers, bullock are less efficient and ovine are also of poor quality. Bullocks are very expansive in relation to input output ratio because these are used for ploughing and thressing only while these may also be used as a carriers of manures, grasses and other materials.

There is a great and urgent need of improving animal breeds and establishing more verterinary centres.

Introduction of farm and social forestry can meet the requirement of quantitative and qualitative fodder.

7.4 INFRASTRUCTURE

Infrastructural facilities are broadly divided into two categories—physical and institutional.

Physical Infrastructure

Mainly the facilities of irrigation, transport and electricity are included in the physical infrastructure which play a significant role for all round development of rural areas.

Irrigation

Overwhelming population in the area derives its livelihood from agriculture. As the upland agriculture depends upon rainfall the crop yields are very low and inhabitants of these villages fulfil their food requirements from market supplies.

The temporary *gules* constructed from seasonal hill streams by the villagers, suffer from siltation and sliding every year and thus, require additional investments for maintenance. It is clear from table 7.8 that more than 75 per cent area of the cultivated land is irrigated in dun floor. Due to undulating topography the construction coste of canal is very high in mountainous part while it is comparatively lower in dun floor.

Table 7.8 Irrigated Area (1985-86)

Unit	*Irrigated* *(ha)*	*(% of the total cultivated area)*
U_1	27.53	22.55
U_2	30.76	100.00
U_3	20.61	14.27
4_4	274.23	100.00
U_5	35.14	6.20
U_6	140.13	78.42
U_7	4.86	1.00
U_8	378.31	100.00
U_9	9.70	34.67
U_{10}	1204.45	89.66
Kotadun	2123.72	60.03

Table 7.9 clears that the total length of canals in Kotadun is 47.85 km out of which only 11.2 km canals are lying in the hilly part while 36.65 km canals lie in the dun floor. The irrigated area is much lower than that of the capacity of canals. It is mainly due to two reasons—firstly, the heads are located near the contact zone of two lithological units of Lesser and Outer Himalaya along M.B.T. and thus, are destroyed by degradational processes or neo-tectonic activities every year; secondly, most of the water of the canals is lossed due to evaporation, percolation and non productive consumption, the amount of this loss is estimated to be nearly 50% in the hills and 50-70% in dun area.

Table 7.9 Distribution of Canal

Canal	*Length km*	*Irrigational Capacity (ha)*
A. Hill:		
1. Amgarhi	1.70	10.00
2. Amota	1.20	81.00
3. Gadela	0.85	6.00
4. Chara	1.25	12.00
5. Basi	3.00	2.00
6. Syat	0.13	under construction
7. Adhaura	3.07	11.00
B. Dun Floor:		
1. Bhalaun	0.80	113.00
2. Patkot	2.20	290.00
3. Hansrampur	0.40	6.00
4. Rampur	1.50	57.00
5. Bhatlani	2.55	106.00
6. Kota System	18.00	1196.00
7. Dechauri System	11.20	995.00
Total	47.85	2888.00

Source: Irrigation Department, Nainital.

Transport: It is an effective means of interaction between villages and central places. From the view point of development of transport net-work, Kotadun is under developed as compared to other parts of Kumaun. Since the transport is the backbone of development, there activities are concentrating near the roads. Contemporary knowledge suggests that accessibility abets the development. That is why the promotion of increased mobility is a primary objectives of development planning every where (Khanka, 1988). Evidently the development of a transport net work is essential need in any region. The hostile physical environment lack of credit and marketing facilities and inadequate transport and communication facilities are responsible for under development of hill region (Katar Singh, 1977). In the hill it is only transport that helps in linking and developing different sectors of economy through the criss crossing (Khanka, 1987). Ramnagar-Kotabagh and Kaladhungi-Kotabagh roads are partly mettled. Nainital-Kunjakharak, Ramnagar-Patkot, Pawalgarh-Garjia roads are unmettled or under construction for the last 10-15 years. There is no regular bus services in these roads. Under these considerable conditions how one can introduce the technological advancements. Majority of the inhabitants keep ponnies for carrying goods otherwise one has to negotiate it by head loads. A few well off families keep tractors in the dun valley. Under new schemes of community development a few bridle roads—Kotabagh-Bajuniahaldu, Dechauri-Pangot, Kotabagh-Ghu-Ghu, Kotabagh-Baghani, Kotabagh-Patkot, Patkot-Banghani, Patkot-Okhaldunga are now maintained but the condition is still bad to worst. Ramnagar (40 km) and Haldwani (40 km) are the nearest railway station and market centres. Due to lackness of transportation facilities, the inhabitants return after two or three days from these centres while the maximum distances of the remotest village do not exceed 50 km from these markets. Table 7.11 and fig. 7.4 reveals that 34% of the total villages lie above 4 km distance from the motorable roads. Under these circumstances inhabitants of remote areas are tending to migrate towards nearly urban centres in search of employment and better facilities.

Electricity: Rural electrification programme should be taken as essential component for the resource development in order to ensure that by the end of seventh five year plan period, 70 to 80 per cent of rural population would be benefited with this facility.

More than 50 per cent of the villages are electrified in the study area and the situation is comparatively better than that of the other parts of Kumaun Himalaya. But it is worth to notice that the villages provided with single light connection are declared as electrified. The connections of electricity are generally provided for domestic purposes and only 57 out of 97 villages have so far been taken under this so called electrification programme (Table 7.10). But there are regional gaps in execution of this programme as none of the villages in U_3 could be electrified while all the villages in U_6 and U_{10} enjoy this facility. The observations reveal that only road side villages have been electrified. Owing to the high transportation charges of poles to the remote villages, these remain still unelectrified. The electricity is supplied from Banbasa.

Table 7.10 Electrified Villages (1986-87)

Unit	*Number*	*% of the total villages*	*% of the unit*
U_1	1	1.03	20.00
U_2	1	1.03	50.00
U_3	–	–	–
U_4	2	2.06	50.00
U_5	12	12.37	54.54
U_6	4	4.13	100.00
U_7	10	10.31	55.55
U_8	10	10.31	90.91
U_9	1	1.03	20.00
U_{10}	16	16.49	100.00
Kotadun	57	58.76	–

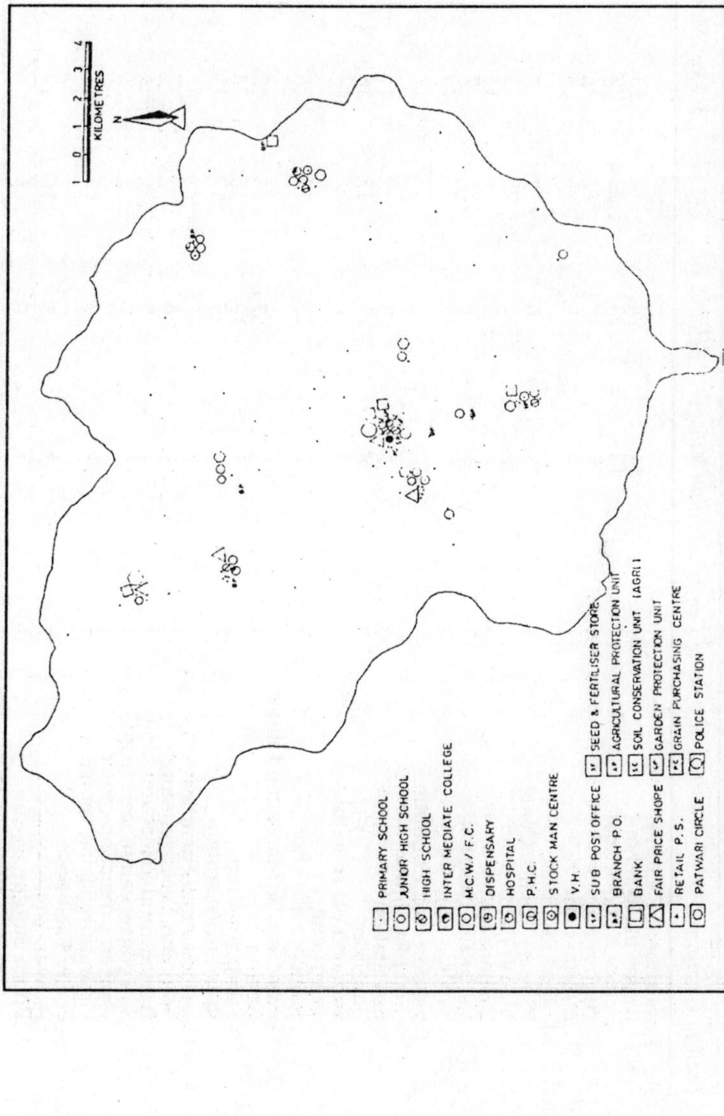

Fig. 7.4 Kotadun: Institutional Infrastructure (1987–88)

Table 7.11 Kotadun: Institutional Infrastructural Facilities (86-87)

Facility Group	*No. of Centre*	*Within Village*	*Below 2 km*	*2-3 km*	*3-4 km*	*More than 4km*	
A. Educational Facilities:							
Junior Basic School	44	44	24	16	13	-	97
Senior Basic School	12	13	23	22	22	19	99
High School	4	3	9	11	12	62	97
Intermediate	1	1	6	7	7	76	97
B. Medical Facilities:							
Primary Health Centre	1	1	6	7	7	76	
MCW/FC	10	10	21	23	15	28	
Dispensary	4	4	5	9	11	68	
Vertinary Hospital	1	1	6	7	7	76	
STMC	5	5	15	17	25	35	
C. Transport and Communicational Facilities:							
Bus Station	2	2	5	8	11	71	
Bus Stops	8	8	21	17	18	33	
Sub post office with Telephone	1	1	6	7	7	76	
Branch post office	7	7	23	13	16	38	
D. Commercial & Marketing Centre:							
Bank	4	3	7	5	10	72	
Fair price shop	9	7	13	16	8	53	
Retail price shop	13	13	25	17	11	31	
E. Developmental Facilities							
Seed and Fertilizer store	1	1	6	7	7	76	
Agricultural Protection Unit	1	1	6	7	7	76	
Soil conservation unit (Agri)	1	1	6	7	7	76	
Garden protection centre	1	1	6	7	7	76	
Grain purchasing centre	1	1	6	67	7	76	
F. Security Facilities:							
Patwari circle	7	7	15	10	19	46	
Police Station	1	1	6	7	7	76	

Source: Concerning departments.

Institutional Infrastructure

The services and amenities of educational, medicinal, communicational, marketing and agricultural extension play pivotal role in the development of rural areas. Backward and remote areas are recognized on the basis of absence of these services and facilities which hinders the advancement of the area and encourages the migration of people towards nearby centres of Nainital, Haldwani, Ramnagar and Kashipur for availling these facilities.

Table 7.11 reveals that there are 44 Junior Basic Schools, 12 Senior Basic Schools, 4 High schools (3 boys & one girls), 1 Inter college, 1 Primary Health Centre. 10 MCW/FC, 4 Dispensaries, 5 Stock Man Centres, 1 Vertinary Hospital, 2 Bus Stations, 8 Bus Stops, 1 Sub Post Office with Telephone, 7 Branch Post Offices, 9 Fair Price Shops, 13 Retail Price Shops, 7 Patwari Circles and one each for Seed and Fertilizer Store, Agricultural Protection Unit, Grain Purchasing Centre and Police Station. Although some mobile units of grain purchase are opened on temporary basis during thressing period in the dun villages. Banking facility is available in 3 centres of the area—Amgarhi RRB, Aonlakot—BOB and DCB and Pangot-RRB.

The analysis reveals that the development of the services and facilities of basic physical and institutional infrastructure is far below from the adequate and more than 60 per cent of the total villages fall more than 4 km away from the location of these services.

Chapter 8
Landuse and Cropping Pattern

The landuse in an area depends upon the objectives of life of the people living therein. It represents the interaction of physical and economic factors (Stamp, 1964). The land is used for cultivation, pastoral activities, afforestation and construction of buildings, roads, railways etc. The type of landuse differs from one place to another owing to its complex environmental factors.

Land classification is very complex problem. It is needed as a basis for landuse planning and in most cases for determining the relative utility of land for different types of farming or for other purposes.

The landuse in Kotadun is rather traditional than scientific. In reference to human occupation people depend upon agriculture, grazing, forestry, part-time quarrying and seasonal labour. Almost all categories of landuse are under deterioration. There are several problems of landuse—decreasing fertility of the soil, deforestation due to increasing demands of land for cultivation and ignored pasture and waste lands.

8.1 CLASSIFICATION OF AREA

The land of Kotadun is broadly divided into two categories:

1. Land owned by State Forest Department or the Reserved Forest Area, and

2. Land owned by State Revenue Department or the Revenue Area.

Table 8.1: Classification of Area (1985-86)

Unit	*Geographical Area (ha)*	*Forest (ha)*	*Area (%)*	*Revenue (ha)*	*Area (%)*
U_1	1,945.00	1,387.47	71.34	557.53	28.66
U_2	3,250.00	3,206.11	98.65	43.89	1.35
U_3	2,565.00	1,929.98	68.62	805.02	31.38
U_4	4,551.00	4,097.07	88.09	553.93	11.91
U_5	6,893.00	2,894.99	42.00	3,998.01	58.00
U_6	1,364.00	1,047.92	76.83	316.08	23.17
U_7	6,994.00	5,751.78	82.24	1,242.22	17.76
U_8	3,027.00	2,244.69	74.16	782.3	25.84
U_9	425.00	380.60	89.55	44.40	10.45
U_{10}	4,964.00	3,266.62	65.81	1697.38	34.19
Kotadun	36,078.00	26,037.23	72.17	10,040.77	27.83

The Geographical area of Kotadun 360.78 km^2 out of which 260.37 km^2 or 72.17 per cent is under the control of the state Forest Department and 100.41 km^2 or 27.83 per cent though owned by state Revenue Department, is under the direct control of the rural society of 96 villages. In fact, such large proportion of forest area is not the actual vegetal cover but it is simply the area under possession of Forest Department. An interpretation of landsat imagery indicates that only 28.7 per cent of Indian Central Himalaya is now forested and only 4.4 per cent of the area has a forest with greater than 60 per cent crown density (Singh *et al.*, 1984). Since the State Forest Department manages and develops the forest land in its own way, the area under this land is not taken into consideration for the present investigation. Land degradation maximally occurs in the civil area because of uncontrolled and unscientific activities of common mass. The area requires intensive care and immediate treatment in order to avert the worst consequences.

8.2 LANDUSE

The landuse of Kotadun is given in table 8.2 and dipicted in figure 8.1. The reporting or the revenue area consists of six category uses forests, pasture, not available for cultivation, orchards, cultivated and culturable waste area.

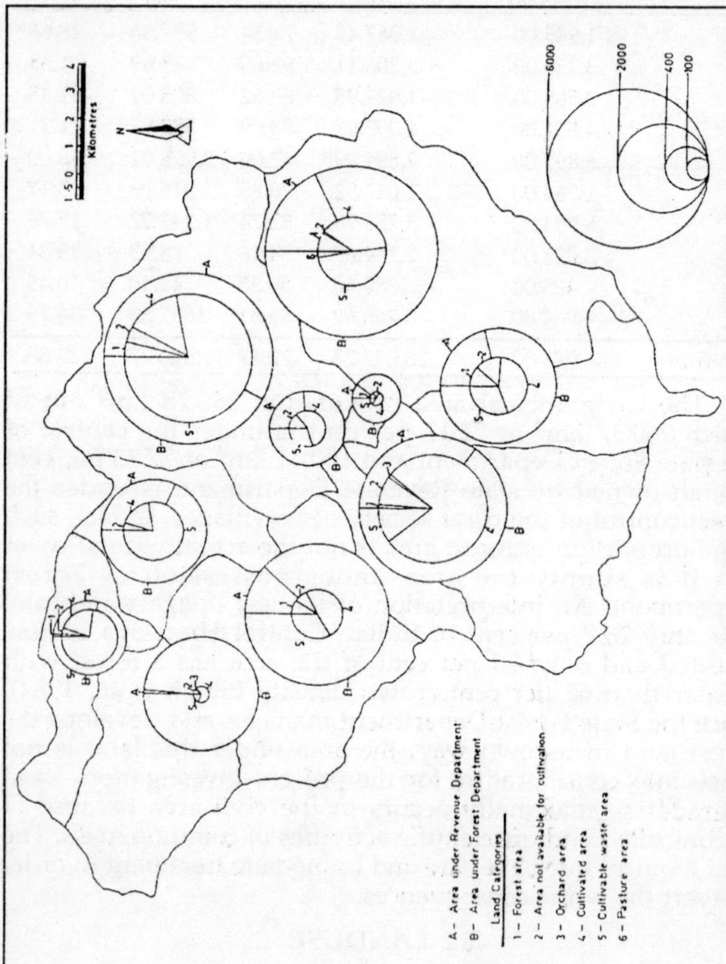

Fig. 8.1 Kotadun: Major Landuse (1985-86)

Table 8.2: Landuse (1985-86)

	Landuse Category	*Total area (ha)*	*% of total*	*% of each category*	
	Geographical area	36,078.00	100.00	..	..
	Reserved Forest area	26,037.23	72.17	..	..
	Revenue or Reporting area	10,040.77	27.83	..	..
	Total	10,040.77	100.00	..	..
A.	Forest area	354.54	3.53	100.00	..
	(i) Panchayati	5.38	..	..	1.52
	ii) Civil	195.39	..	..	55.11
	(iii) Shrubs and bushes	153.77	..	..	43.37
B.	Pasture area	38.18	0.38	100.00	..
C.	Area not available for cultivation	586.37	5.84	100.00	..
	(i) Settlements	76.08	..	..	12.97
	(ii) Roads and Paths	63.29	..	..	10.79
	(iii) Water bodies	401.18	..	..	68.42
	(a) Natural	(311.17)	..	..	(53.07)
	(b) Cultural	(91.07)	..	..	(15.35)
	(iv) Barren	45.82	..	..	7.82
D.	Culturable waste area	5,496.40	54.74	100.00	..
	(i) Waste	5,218.14	..	..	94.94
	(iii) Fallow	278.26	..	..	5.06
	(a) Old	(185.08)	..	..	(3.37)
	(b) Current	(93.18)	..	..	(1.69)
E.	Orchards	27.56	0.28	100.00	..
F.	Cultivated area	3,537.72	35.23	100.00	..
	(i) Irrigated	2,123.72	..	..	60.03
	(ii) Unirrigated	1,524.00	..	..	39.97
	(a) Rabi net cropped	(3,123.65)	..	..	(88.30)
	(b) Kharif net cropped	(3,375.49)	..	..	(95.41)
	(c) Double cropped	(2,961.42)	..	..	(83.71)
	(d) Gross cropped	(6,499.14)	..	..	(183.71)

Forest Area

The forest is one of the major forms of natural landscape. Climate, soil and surface features play a key role in determining the type, abundance and variety of vegetation. The high temperature and humidity produce a climatic optimum for luxurient vegetative growth in Siwaliks and Dun floor. On steeper slopes, along drainage channals and other places where light can penetrate to the forest floor a dense growth of lower trees, shrubs and varies take command. The seasonality of rainfall is reflected by the forest vista. A monotanous brown vista during dry season is replaced by pleasing successions of green during the rains.

Maximum area of Kotadun suffers from several forest problems. Expanding cultivation and construction have encroached upon the forested area now covering only 3.53 per cent (354.53 ha) of the reporting area. Forests are classified as reserved, civil and panchayati on the basis of ownership and legal status. The reserved forests are owned and managed by the state Forest Department. The civil and panchayati forests including shrubs and bushes (community forests) are owned by the village community but jointly administered by the state Revenue Department, Van Panchayat and Gram Sabha. The private forests are owned and managed by individuals. In Kotadun the forest area includes panchayati forest (1.52%), civil forest (55.11%) and shrubs and bushes (43.37%). Today one of the most considerable problem of soil conservation lies on afforestation of the river basins. It is more alarming along the Main Boundary Thrust (MBT) and in the Siwalik zone. According to Champion as quoted by Day (1961), during the half century one serious setback can not be overlooked as it will be so difficult if indeed possible to remedy. It is the withdrawal of the protection of the community forest from the Forest Department and resultant degradation of these forests and even contagious reserved forests.

The forest wealth of the Himalaya is vital for regional prosperity along with soil conservation associated with the Gangetic plain. Heavy rainfall, steep and exposed slopes, ruthless deforestation, excessive grazing and injudicious cultivation on slopes have caused serious problems of erosion. Land degradational processes are very common along MBT and Siwalik foot-hills. These have led the burning problems of siltation of low lying areas. Erosion and siltation may have little adverse affect for this area itself, but are fall vital significance to the plains below 400 m elevation (Sinha, 1961). For the environmental restoration requirements these areas are best suitable for forest and cannot be used for agricultural and other purposes. These are required conservation and development on their own way.

Pasture Land

Pasture land occupies 0.38 per cent (38.18 ha) of the total reporting area. There is no demarcation between the pasture and the waste land. Forests and waste land are generally used for grazing. The area lying between the cultivated land and the reserved forest may be include in pasture land. After continuous exploitation and grazing, these areas now may be seen as waste land. Their treatment lies on cultivation of local grasses both for soil conservation and grassland development.

The seasonal pastures in the Siwalik and Dun floor are spread under the reserved forest. There are some facilities of grazing and living for animals and caretakers—availability of flat land, drinking water, grasses and fodder. Those places used for seasonal pastoral activities are known as 'KHATTA'.

Combination of grazing and grass lands are the necessary features of agricultural economy. Development and management of these lands are very indispensable requirements of landuse planning. Cattle may graze freely in the forest areas as they are not supposed to harm the forest lands. Notwithstanding goats and sheep are more selective feeders yet they are harmful for forest grazing and brauzing due to their easy access to trees.

Some of the land under pasture is termed as civil unmeasured (*benap*) land, which is generally bare. Civil land and Panchayati forest are at the disposal of villagers and thus, are exploited unscientifically. As a result, these convert into waste lands which are neither used for growing crops nor used as rich grazing lands. From view point of nutritional value of the grasses found in the area is very low.

Improper management of pastures results in damage the forest lands and degradation of the soil. The deterioration of the forest soil which brings with it the degradation both of the forest lands and forage resources, is perhaps the worst in civil due to overgrazing (Leopold, 1953). On account of sparse pastures in Kotadun, fodder bearing plant species are required to be regenerated here for two fold benefits of providing fuel to the villagers and fodders to the animals. Various fodder grasses should also be introduced for the management of pasture and waste lands.

Area Not Available for Cultivation

This is unproducing area under settlements, roads, pedestrians, water bodies and barren lands. Total area under this category is 586.37 ha or 5.84 per cent of the reporting area.

Settlement: The settlements cover 12.97 per cent (76.08 ha) of the area not available for cultivation, it includes area under houses, shops, courtyards, cattlesheds, temple, water mills, funeral places etc. Settlements are established on unproducing land usually upon the rock bases out-cropping above the adjoining lands. Courtyards are common features infront of the houses, shops and cattlesheds, connected by pedestrians. The ground floor of the houses are generally used for keeping the animals and where the number of animals is larger, separate sheds are made. Water mills are located either on river banks or on village canals. The funeral places are situated at a distance from settlements near a perennial stream.

Roads and Paths: This category land includes area under motor roads, cart tracks, bridle paths etc. Area under this category is 10.79 per cent (63.29 ha) of the area not available for cultivation. Till 1970, the area was approached by cart tracks and bridle roads but during a short period of fifteen years or so, a change has occurred due to the construction of motor roads owing to the consciousness of balanced rural development.

Water Bodies: This category of land is comprised of 68.42 per cent (401.18 ha) of the area not available for cultivation. It can further be divided into two sub-categories, Natural (3.07%) and cultural 15.35 per cent. The natural water bodies include area under streams, springs etc. While the cultural water bodies include that under canals, guls, reservoirs etc.

Barren Land: This category land covers small portion of the area. It is comprised of 7.82 per cent of the area not available for cultivation. It includes area under boulders, escarpments and otherwise barren lands.

Orchards

From the environmental point of view, the area is much appropriate for horticulture. In many parts farmers can get more profitable returns from orchards than raising seasonal crops. It is suitable for subtropical as well as temperate fruits. The fruits which are grown here are mangoes, guavas, jackfruits, apricot, pears, apple, citrus fruits, strawberies, blackberries, plantains, papaya etc. This category land covers 0.28 per cent of the reporting area.

Although every unit possess a little proportion of area under this category, but there is not a definite fruit belt in the region. The upper basins of Dabka and Baur rivers produce lemons, oranges, apricots, pears, strawberries etc. While their lower portions of almost plain areas, are famous for mangoes, leeches, guavas, jackfruits, plantains, blackberries, papaya etc. There is a great need of development of orchards in the area from view point of economic development as well betterment of landuse.

Culturable Waste Land

The culturable waste land encompasses 54.74 per cent (5496.40 ha) of the reporting area. This land may have been used for cultivation but being relatively less capable for cultivation, it remains waste in the out skirts of the cultivated fields and is often used for grazing. It includes waste and fallow lands. In Kotadun highest percentage of reporting area comes under this category.

Waste Land: Total area of waste land is 94.94 per cent of the culturable waste land. Due to low fertility and little production, this category land remains as permanent fallow. The waste land includes degraded land patches lying fallow around or in between cultivated fields. Small land patches, with a little texation, are under the ownership of cultivators but due to their less productive nature by one reason or the other, these are left as permanent fallow. Such type of waste land can be used for horticulture, farm forestry and growing rich nutritive fodder grasses. Moreover, vast land patches of public waste land occurring on account of diverse topography, weak geology, climatic variations and biological factors, are affected by accelerated soil erosion, landslides and mass wasting. Restoration of these degraded land patches is the most urgent requirement of the time. There may be put under introduction and development of social forestry.

Fallow Land: The arable land which has not been cultivated for a few years, is known as fallows. This category land comprises 5.06 per cent of the total culturable waste area. It can further be divided into two sub-classes-old (3.37 per cent) and current (1.69 per cent). Old fallow includes the area which has not been cultivated for part two or three years. The land which is not used for raising crops during the reporting year, is generally known as current fallow. Fallow land percentage is very low due to shortage of arable land. Along with scientific preferments and explosion of population, peasants are bent upon keeping the maximum possible land under plough. The principal reason of this following practices is to restore the lost soil fertility over a period of centuries. But it can be compensated by heavy manuring and application of proper crop rotations.

The distribution of culturable waste area is just opposite to cultivated area. There is an antipodal arrangement between the cultivated area and cultivable waste area in Kotadun. This is so because of the lackness of attention towards proper utilization of land in these areas where the land resource is adequate and the pressure of population is not so heavy.

Keeping in view the future prospect, scientific planning and development of the waste land for optimal use is most urgent within the permissible limit of environmental conditions.

Cultivated Area

Cultivation is the most important and man's oldest occupation. The physical factors of topography, drainage, climate and soil play important roles in determining extent of the cultivated area. It covers 35.23 per cent of the reporting area in Kotadun. Table 8.3 clearly idicates that the highest percentage in lower units is due to the favourable environmental conditions of plain topography, fertile soil, better irrigation facilities etc.

There are two specific agricultural regions; firstly, the elevated fluvial terraces and secondly, the sloping areas of '*Talaon*' and '*Upraon*' lands. The *Talaon* land is spread over the dunfloor lying below 900 m elevation zone. This area is privileged with good quality of soils and facilities of irrigation and transportation and thus, is capable of producing two or three crops a year. The Upraon land, on the other hand, requires special attention in building and maintaining the terraces for cultivation. The scarcity of irrigation facility is the only reason for dry farming in this land.

Kotadun has remarkable probabilities for agricultural development provided that scientific methods of cultivation are practised here. It has got ample range of temperature, rainfall and soils. It will be almost obvious that the finest arable soils are these of as intermediate texture with an admixture of sand, silt and clay particles, in other words, the best arable soils are the loams. The combined relative case of mechanised working with facility of retaining moisture and manures are because of these relative qualities are often naturally rich in plant foods

Table 8.3 Unit wise Distribution of Landuse (1985–86)

Unit	Reporting area	Forest area		Area not available for cultivation		Orchards		Culturable Waste area		Cultivated area	
	(ha)	(ha)	(%)	(ha)	(%)	(ha)	(%)	(ha)	(%)	(ha)	(%)
U_1	557.53	9.72	1.74	17.49	3.14	0.81	0.15	407.44	73.08	122.07	21.89
U_2	43.89	–	–	4.70	10.70	0.81	1.85	7.62	17.36	30.76	70.08
U_3	805.02	–	–	31.51	3.91	0.11	0.02	628.94	78.13	144.46	17.94
U_4	553.93	47.94	8.65	45.68	8.25	0.69	0.12	185.39	33.47	274.23	49.51
U_5	3,998.01	134.49	3.36	163.10	4.08	2.56	0.06	3138.10	78.50	559.76	14.00
U_6	316.08	–	–	45.08	14.26	0.72	0.23	91.58	28.97	178.70	56.54
U_7	1,242.22	53.52	4.31	52.85	4.25	2.53	0.20	611.27	49.21	483.87	38.95
U_8	782.31	88.34	11.29	73.43	9.89	0.87	0.11	241.36	30.85	378.31	48.36
U_9	44.40	7.08	15.95	7.47	16.82	0.57	1.28	7.07	15.42	22.21	50.02
U_{10}	1,697.38	13.45	0.79	145.06	8.55	17.89	1.05	177.63	10.46	1,343.35	79.15
Kotadun	10,040.77	354.54	3.53	586.37	5.84	27.56	0.28	5,496.40	54.74	3,537.72	35.15

The southern aspects viz. south facing slopes get good amount of sunshine and precipitation and thus, these are in suitable conditions for cultivation. The northern aspects are mostly covered with dense vegetation. The sites usually selected for terracing are the sunny slopes which often have a southern aspects. In these sites the sun's rays fall almost at right angle upon the soil and the consequences of this district sunlight, snow and frost disappear quickly.

Double cropped area means area sown more than once a year. Due to the scarcity of cultivated land in reference to fulfil the food requirements of ever increasing population in the region, 83.71 per cent area is under double cropped while in the cases of land near the settlements and is good quality lands more than two crops in a year. It is also observed that soyabean, lahi and wheat are taken in one year.

Irrigated area is only that land which is irrigated by the means of canals. Besides this some upland areas are irrigated by seasonal gules leading from the seasonal streams. These gules are silted every year and require additional investments for their maintenance. Otherwise the agriculture always depends upon the rainfall and the crop yields are very low. These areas, thus, depend upon the imported food grains, as the quantity of field produce is only for two or three months.

Since 60.03 per cent of the cultivated area is irrigated by canals, 57.32 per cent irrigated area lies in lower units of dunfloor and rest 2.71 per cent area is in upper units of mountainous part. The main reason of these differences is the uneven topography due to which the cost of canal construction reaches very high in compared to the level areas. It is also observed that the site of canal heads are wrongly selected from the geological point of view, near MBT which is very sensitive zone proning to neo-tectonic activities.

8.3 CROPPING PATTERN

Cropping pattern denotes to the proportion of the area given to different crops in a particular crop season. A change in cropping pattern signifies a change in a proportion of the area under different crops. Cropping pattern of an area depends upon physical characteristics particularly soil and climatic as well as social systems in general.

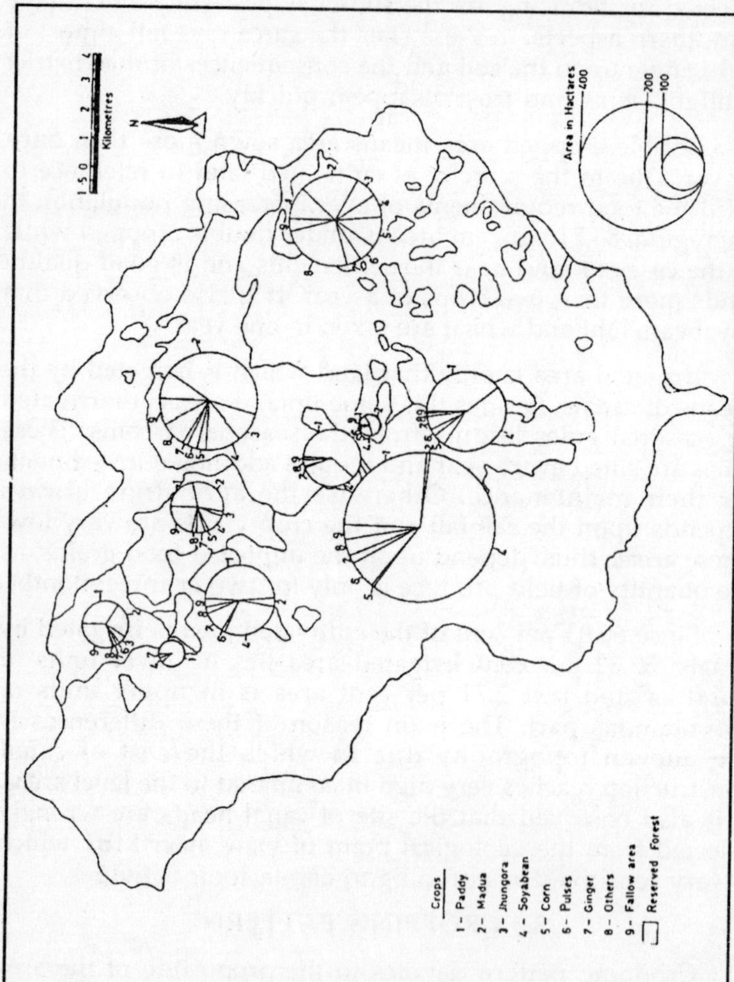

Fig. 8.2: Kotadun: Kharif Landuse (1985)

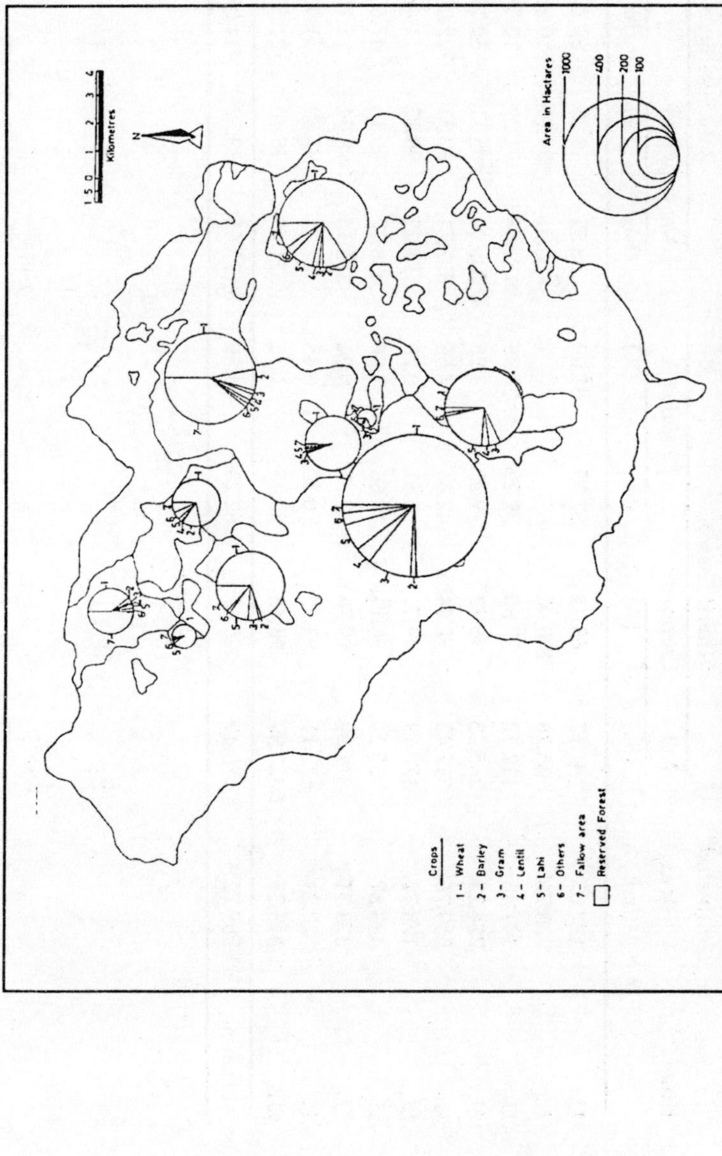

Fig. 8.3 Kotadun: Rabi Landuse (1985-86)

Table 8.4: Cropped Area (1985-86)

Unit	Cultivated Area	Kharif Season				Rabi Season			
		Net Cropped		Fallow		Net Cropped		Fallow	
	(ha)	(ha)	(%)	(ha)	(%)	(ha)	(%)	(ha)	(%)
U_1	122.07	84.12	68.91	37.95	31.09	58.87	48.23	63.20	51.77
U_2	30.76	36.76	100.00	–	–	30.51	99.19	0.25	0.81
U_3	144.46	112.17	77.65	32.29	22.35	131.88	91.29	12.58	8.71
U_4	274.23	265.23	96.72	9.00	3.28	235.04	85.71	39.19	14.29
U_5	559.76	517.42	92.44	42.34	7.56	337.71	60.33	222.05	39.67
U_6	178.70	178.19	99.71	0.51	0.29	177.22	99.17	1.48	0.83
U_7	483.87	465.21	96.14	18.66	3.86	436.27	90.16	47.60	9.84
U_8	378.31	377.39	99.76	0.92	0.24	373.15	98.64	5.16	1.36
U_9	22.21	22.11	99.55	0.10	0.45	21.21	95.50	1.00	4.50
U_{10}	1,343.35	1,322.89	98.48	20.46	1.52	1,321.79	98.40	21.56	1.64
Kotadun	3,537.72	3,375.49	95.41	162.23	4.59	3,123.65	88.30	414.07	11.70

Table 8.5 Cropping Pattern (1985-86)

Particulars	*Area (ha)*	*% of cultivated*	*% of net cropped*
Paddy	955.62	27.01	28.31
Madua	470.75	13.32	13.95
Jhungora	98.83	2.79	2.93
Corn	94.77	2.68	2.81
Soyabean	1,230.83	34.79	36.46
Pulses	152.26	4.30	4.51
Ginger	129.60	3.66	3.84
Others	242.83	6.86	7.19
Total Kharif Cropped Area	3,375.49	95.41	100.00
Fallow area	162.23	4.59	..
Wheat	2,414.76	68.20	77.30
Barley	128.91	3.64	4.13
Gram	171.56	4.85	5.49
Lentil	103.27	2.93	3.31
Lahi	234.94	6.64	7.52
Others	70.21	1.98	2.25
Total Rabi Cropped Area	3,123.65	88.30	100.00
Fallow area	414.07	11.70	..
Total cultivated area	3537.72	100.00	..

Table 8.6: Unit wise Distribution of Various Crops (1985-86) (*Percentage of the cultivated area*)

Unit	Kharif Crops						Rabi Crops					
	Paddy	Madua	Jhungora	Corn	Soyabean	Others	Wheat	Barley	Gram	Lentil	Lahi	Others
U_1	33.78	25.20	5.30	–	–	0.66	40.42	2.65	0.33	0.65	0.99	3.19
U_2	86.86	–	–	–	6.57	–	93.43	–	-	-	5.27	0.49
U_3	21.21	32.15	2.80	1.96	11.08	2.48	72.23	12.50	-	1.47	2.28	2.81
U_4	48.81	3.73	–	–	29.36	3.08	70.80	1.24	4.43	0.40	7.36	1.48
U_5	9.38	45.52	9.18	2.75	12.06	8.13	46.97	7.81	0.32	2.33	0.91	1.99
U_6	61.67	–	–	–	32.29	3.67	95.44	4.43	0.68	0.51	0.72	0.39
U_7	12.06	20.98	7.63	10.38	8.31	25.00	66.55	9.00	1.24	2.92	8.25	0.26
U_8	39.69	–	–	5.11	44.94	1.35	67.67	–	5.55	3.48	20.95	0.69
U_9	15.80	55.43	–	2.03	22.50	0.32	68.84	13.51	5.09	3.38	3.74	0.94
U_{10}	25.94	1.09	–	0.49	58.88	3.57	74.79	1.27	9.52	4.26	6.12	2.44
Kotadun	27.01	13.31	2.79	2.68	34.79	6.84	68.26	3.72	4.85	2.92	6.64	1.98

As it is clear from the table 8.4, the cultivated area covers 35.23 per cent (3537.72 ha) of the reporting area. The area represents different cropping pattern which basically is an outcome of the varying physical environmental features and subsistence nature of agriculture. The study of cropping pattern is based on seasonal landuse.

Kharif Landuse

Generally the kharif season is the period between June to October. Table 8.4 reveals the kharif landuse and cropping pattern of the area. Kharif is an important agricultural season of the area where 95.41 per cent of the cultivated area is sown in this season. Table 8.5 and fig. 8.2 portrays the distribution of various crops in the kharif season. This is only the season when all the possible land of the area may be cultivated. Table 8.5 also reveals that the 4.59 per cent (162.23 ha) is left as fallow.

Kharif Cropping Pattern: About 48.00 per cent of the net cropped land is devoted to cereals-paddy, madua, jhungora and corn and 40.97 per cent to pulses,—soyabean, urd, gahat, bhat etc. Remaining 11.03 per cent area is devoted to non cereal crops-ginger and vegetables. Table 8.5 and fig. 8.2 shows the kharif cropping pattern of the region.

Rabi Landuse

Rabi crop season is the period between October to April and May. Table 8.5 reveals that the rabi landuse and cropping pattern of the area. Agriculturally, rabi season is considered as a most important season and 88.30 per cent (3223.65 ha) of the cultivated area is devoted to the cultivation of rabi crops. It is clear from the table 8.5 & fig. 8.3 that 11.70 per cent (414.07 ha) area is left as fallow owing to infertility of the soil.

Rabi Cropping Pattern: Net 81.43 per cent (71.90 ha) of the net cropped area is devoted to wheat and barley. Remaining 18.57 per cent area is devoted to gram, lentil, lahi and other crops (Table 8.5 & fig. 8.3).

Chapter—9

Agro-climatic Zones, and Sampling Framework

9.1 AGRO-CLIMATIC ZONE

Of all the environmental factors, altitudinal variation plays dominant role in determining the landuse. The altitudinal range in Kotadun varies from 400 to 2600 ma.s.l. while some of the peaks of Kunjakharak-Binayak range rise above 2600 m (Badhantoli 2623 m) and those of Babilia-Kailkhura range above 700 m. The climatic conditions in Kotadun vary in order to altitude. Climate and relief are the most important influencing factors for determining landuse in general and agriculture in particular.

An attempt has been made to analyse the impact of climate and altitude on landuse. Kotadun is divided into three identified major altitudinal agro-climatic zones. Another aim of the study is to suggest the cropping pattern for improving the agricultural situation.

Human activities depend on the physical as well as the cultural environment under the limitations of which one lives and to some extent tries to minimise these limitations. Visher (1932) has rightly put forward that the special field of geography concerns itself with the study of influences of natural environment on the nature and distribution of human activities. Right from the beginning works in the field were attracted to the problems of explaining how variations in environment influenced the agricultural landscape (Gregor, 1970), McHary (1966 & 1969) emphasized the importance of man's harmony with nature and the need of understanding and coordinating the natural resources with those of human being. Following these principles distinguished as ecological determinism, he has shown how nature imposes limits, provides guidelines and trains man in solving environmental problems. Nature in its diverse manifestation, namely, the soil, water and climate, provides, man indifferent areas with a variety of possibilities of development (Hettner, 1947). Certainly, no other branch of human geography deals so directly with physical environment as Agricultural Geography does (Singh and Dhillon, 1984).

It is necessary for geographical investigation to identify the basic regional differences in agricultural formations also to understand the resultant agricultural type and cropping pattern. Agricultural complexes vary from place to place due to the relief, soil, climate etc. Keeping in view the existing cropping pattern and future development possibilities, the schemes are to be formulated in the light of prevailing natural and social conditions.

Agricultural Geography is not only the study of crops but it also forms an applied ecology: Animal husbandry is also an integral part of agricultural geography. Both the raising of crops and animal husbandry depend on the limitations of natural as well as cultural environment. Therefore, it is the systematic way to study the physical conditions influencing the pattern of agricultural landuse.

The area is divided into three altitudinal zones below 900 m, 900-1800 and above 1800 m. Climatic data of Aonlakot represent the zone lying below 900m, and those of Nalni represent the intermediate zone between 900-1800 m. The climatic data of Sigarhi represent the third altitudinal zone rising above 1800 m. Due to unavailability of the micro-level climatic data, the area is divided into three major zones only.

Altitude

Among the terrain, altitude plays a significant role for utilization of land. As a matter of fact, it decreases the air pressure with increase in elevation and *vise versa*. Modifications in temperature do not occur frequently in humid conditions associated with increasing elevation making these areas inhabitable (Klages, 1958) for farming communities. Like air pressure, temperature also decreases with altitude owing to the rarefied air having low content of carbondioxide, moisture and other particles. It means it is not that the decreasing temperature with altitude in every region but it depends on other factors of slope angle and sunlight conditions. The distribution of precipitation also depends on altitude, while the precipitation increases successively from lower to higher places. Since the capacity of air pressure to hold moisture varies directly with its temperature and as this decrease, adiabatitally in ascending air, high lands have heavy precipitation (Symons, 1967).

The study reveals that the dun floor is endowed with intensive agricultural practices, high population density, marginal forest area and well developed infrastructure in relation to hilly tract.

Climate

Climate is another most influencing factor to determine the landuse and other activities of man in a region. Among all the environmental influences to which the man is subjected, climate seems to be the most potent one. In a large scale, climate determines where man may live and thrive, what crop he may raise, what type of home he may appropriately build, what sorts of clothing he may wear and what pests and diseases he must combate (Whiteback, 1932). It is clear that climate allows the range of crops which a country economically produce. This inturn in sets the range of commodities which that country must import if it wishes its people to live a full life in the modern sense (Stamp, 1964). At the climate of countries changes, so will its plants or crops (Thomas, 1967). Raising the crops do not depend upon the limited variables but it is controlled by the various elements (Temperature, air pressure system, air masses, winds, snow, humidity, sun shine, rainfall etc.) acting in combination at a time. All these variables fluctuate from time to time, place to place and even season to season.

Owing to all these effects, the sub-tropical climate in dun floor encourages the cultivation of wheat, paddy, soyabean, peas etc. while the temperate climate on higher places allows to grow the millets. Besides cropping pattern, other landuse processes are also affected by altitude as the share of grazing area is higher on the hills as compared with the dun floor. A variety of fruit culture is also determined by the climate. Mangoes, leechi, guavas, etc. are generally grown in the dun floor while nuts, oranges, lemons, strawberries, mulberries etc. are raised in the hilly region. Due to extremely cold climate on higher elevations of Kotadun generally the land is left as waste while every patch of land is occupied by crops in dun floor. Owing to these factors, the grain crops could not be grown on higher places particularly in kharif season.

On the basis of above reasons (Table-9.1) the study area is divided into three altitudinal agro-climatic zones.

1. The first Altitudinal Agro-Climatic zone.
2. The second Altitudinal Agro-Climatic zone.
3. The third Altitudinal Agro-Climatic zone.

The First Altitudinal Agro-Climatic Zone: The first altitudinal agro-climatic zone (Dun floor) lies below 900 m from sea level. Encompassing an area of 196.47 km^2 (54.45% of total area), it inhabits 38 villages. The average annual temperature is 23.12^{o}c and average annual rainfall is 173.69

cm. Thus, the climate of sub-tropical continental monsoon type is found in the dun floor and lower valleys. Major crops viz. paddy, soyabean, wheat, ginger, peas, lahi, gram, pegionpea and fruits viz. banana, jackfruits, leeches, mangoes etc. are widely grown there. There the environmental conditions are favourable for these crops and the production of improved variety grains is very high. The irrigational as well as institutional facilities are also available there. Due to formation of depositional process, the soils of the dun floor are dilluvial containing high humas content. Modern agro-technological measures tractors, threshers, H.Y.V. seeds, chemical fertilizers, pesticides, anticides etc. are also used there. Maximum proportion of the cultivated land is devoted to cash crops as pulses and soyabean because their labour and irrigation requirement is comparatively less than that of paddy and wheat.

The Second Altitudinal Agro-Climatic Zone: Lying between 900 to 1800 m above sea level the second altitudinal agro-climatic zone (mid slope or intermittent area) encompasses 134.74 km^2 or 37.35 per cent of the total area. It comprises 57 villages. The average annual temperature and rainfall are 16.77°c and 180.25 cm, respectively. On the basis of these conditions this is characterized by temperate continental monsoon type climate. The main crops—madua, jhungora, barley, corn, wheat, unirrigated paddy, ginger, gahat, bhat, lentil, mustard and fruits (lemon, nuts, mangoes, etc.) are generally grown there. Due to the undulating topographic features and associated environmental factors the zone is suitable only for growing the millets and coarse grains. The Main Boundary Thrust (MBT) passes through this zone and makes it very unstable and fragile to accelerate soil erosion and mass wasting problems. Owing to these factors there is a continuous trend of migration towards stable dunfloor or more developed centres in search of better jobs and facilities. The inhabitants of all the 57 villages situated in this

Table 9.1: Altitudinal Agro-Climatic Zones

Altitudinal zone (m) (Region)	*Area*		*No. of Inhabited Villages*	*Average Temp.*	*Annual Rainfall*	*Climate*	*Crops*
	(km²)	*(%)*		*(°c)*	*(cm)*		
Below—900 m (Dun floor or valleys)	196.47	54.45	38	23.12	173.69	Subptropical continental monsoon type	Paddy, soyabean wheat, ginger, peas lahi, gram, pegion-pea, mangoes, leeche, jackfruits, banana, genera, etc.
900-1800 m (Mid slope or Intermittent zone	134.74	37.35	57	16.77	180.25	Temperate continental monsoon type	Madua, jhungora, barley, wheat, paddy, ginger, gahat, bhat, lentil, mustered, lemon, nuts, mangoes, etc.
Above 1800 m (Uplands or Mountains tops)	29.57	8.20	1	15.12	193.63	Cold temper-ate continen-tal monsoon type	Barley, wheat, corn, potato, peas, pears, apple, strawberries, lemon, orange etc.
Total	360.78	100.00	96	–	–	–	–

hazardous zone, are living in the conditions of life and death since any calamity may occur at any time particularly during rainy season. Two major events of August, 1981 and September, 1986 had already happened at Khairar and Gadela where besides lands some people and animals are burried under land slide debris (Jalal, 1985 and Pant *et al.*, 1989). The environmental conditions of this zone are suitable only for producing fruits and developing social forestry.

The Third Altitudinal Agro-Climatic Zone: The third altitudinal agro climatic zone (uplands or mountain tops) lies between the elevation of 1800 m to 2623 m in the Kotadun. Encompassing 29.57 km^2 or 8.20 per cent of the total area of Kotadun, this zone includes only one village. The average annual temperature and rainfall are 15.12° C and 193.63 cm respectively. With the increasing trend of elevation the temperature decreases but the rainfall increases in the area. The climate of this zone is cold temperate continental monsoon type. Due to extremely cold only barley and wheat foodgrains are grown there while potato, peas, apple, strawberries, lemon and orange are grown adequately. Most of the area in this zone is covered with dense oak forests while steep slopes are without soil and vegetal cover. The inhabitants of this zone depend on market supplies of food grains. The transport facility is not available there in this zone and the goods are carried on horse backs or one has to negotiate on head loads. The physical environment of this zone is suitable for developing orchards. There are possibilities of developing the animal husbandry. The panoramic view of this zone is favourable for development of tourism.

9.2 SELECTION OF SAMPLE VILLAGES

There are many sampling techniques for landuse studies. Opinion may differ, however random or probability sampling, systematic sampling and purposive sampling are the important ones.

Based on geology, altitude, climate, soil and general landuse Kotadun is divided into three homogeneous micro-regions. Altitudinal variation plays dominant role in determining the landuse. Thus, the area is divided into three physical strata viz.

(1) Dun-floor or valley zone lying below 900 m above m. s. l. is almost dilluvial plain. Since Patkot is the largest village in this zone it has been selected as the representative of it.

(2) Intermittent or mid slope zone lying from 900 m to 1800 m above m. s. l. is occupied by sand and sandy loam soils. It lies in the transitional zone of the Lower Siwalik and the Lesser Himalaya along the MBT. It not only represents to the most faulty landuse and unscientific management but also to the sensitive, weak, fragile and landslide prone area. Village Amgarhi is selected for representing the landuse of this stratum.

(3) Upland or mountain tops lying above 1800 m above m. s. l. represent the landuse of high altitudinal zone and remote villages. A different type of landuse is seen in these villages. Sigarhi is selected for representing this zone.

9.3 LAND CAPABILITY CLASSIFICATION

A new approach for land capability classification is tried by Pant *et al.* (1988) on the basis of soil depth, soil texture, natural slope, erosion intensity, water holding capacity, sun light condition, quantity of irrigation in a year, accessibility from human settlements and potential landuse (Table 9.2).

There are three main land capability classes, sub divided into six land capability subclasses namely, good quality land A_1 and A_2, medium quality land B_1 and B_2 and poor quality land C_1 and C_2. Table 9.2 makes it possible to determine the land capability of a particular field. The first and foremost task is to ascertain the major land category, namely, good, medium and poor quality of the land of a field.

Table 9.2: Land Capability Classification 'MODEL' for Himalayan Region (Pant *et al.* 1988)

Land capability class	*Variables*								*Potential landuse (PL)*
	Soil depth (1)	*Soil texture* (2)	*Natural slope* (3)	*Erosion intensity* (4)	*Water holding capability* (5)	*Sunlight situation* (6)	*Irrigation condition* (7)	*Accessibility from settlement* (8)	
Good land	Deep (>30 cm)	Loam	Low 10°	Low	High	Sunny	Fully	Low (0-1/2 km)	Intensive cultivation with vegetable and multi cropping.
Medium land	Moderate (15-30 cm)	Graval	Moderate (10-20°)	Moderate	Moderate	Partially	Partially	Medium (1/2 to 1 km)	Extensive cultivation with double cropping and horticulture development.
Poor land	Shallow (< 15 cm)	Stony	High (>20°)	High	Low	Shaddy	Nill	High (> 1 km)	Poor cultivations with fuel and fodder.

Phase 1

Eight variables (Table 9.2) for each quality land to determine the capability of a particular land or field have been taken in the first phase. The land category is determined in following order: (1) Maximum of 150 score points are allotted to good quality land; and thus each of the variables gets a maximum of 18.75 points in this category (2) In the same manner the medium quality land is allotted a maximum of 100 points with 12.5 points for each of the variables. (3) Only 50 points are allotted to the poor quality land. Each of the parameters in this class thus, has 6.25 points.

Phase II

In the second phase the points (0-150) are allotted to each field by grouping method to determine its category. The good, medium and poor quality lands are represented as A, B and C class respectively.

The good quality (A) land exists in between the range of 100-150 score points. This group is further sub-divided into two parts, namely, 126 to 150 and 101 to 125 score points and are represented as A_1 and A_2 respectively. The medium quality (B) land is also sub-divided into 76 to 100 and 51 to 75 score points, denoting B_1 and B_2 categories of medium quality land respectively. Similarly, the poor quality (C) land is sub-divided into 26-50 and 0-25 score points for determining C_1 and C_2 categories respectively.

A perusal of table 9.3 reveals that about 32.97 per cent area of dunfloor villages is under good quality land in which 11.55 per cent area is devoted to vegetables and multi-cropping. Almost 21.39 per cent area of this land is given to intensive agriculture or double and triple cropping. Net 61.99 per cent of the village land is of medium quality land, out of which 37.42 per cent area is in B_1 devoted to double cropping-soyabean in kharif, wheat, lahi and gram in rabi season. Remaining 24.55 per cent area of it is B_2 and which is devoted to soyabean and lahi. Only 5.04 per cent of the total area is under C_1 which is occupied by stream beds or otherwise barren lands (Fig. 9.6).

Table 9.3: Area Under Different Categories of Land in the Sample Villages (1986-87)

Land Category Type		Dun Floor (Patkot) ac	%	Mid Slope (Amgarhi) ac	%	Uplands (Sigarhi) ac	%
Good	A_1	103.83	11.58	12.08	4.64	–	–
	A_2	191.71	21.39	25.84	9.92	–	–
Medium	B_1	335.44	37.42	11.37	4.36	–	–
	B_2	220.25	24.57	34.37	13.20	35.02	14.70
Poor	C_1	45.15	5.04	64.17	24.63	151.60	62.05
	C_2	–	–	112.67	43.25	56.78	23.25
Total	–	896.38	100.00	260.50	100.00	244.30	100.00

The area of mid slope or intermittent village (Amgarhi) is categorised into three main and six sub-categories. Out of the total 14.56 per cent area is under good quality (A_1 and A_2) characterized by sandy loam soil, double cropping and irrigation facilities. About 17.56 per cent area is medium quality (B_1 and B_2) land used for growing madua and wheat. Maximum proportion (67.88%) is under poor quality (C_1 and C_2) land. The soil genesis is poor quality land caused by cressing of rocks along MBT, is not capable of raising crops but is suitable for developing social forestry. It is provided to millets but better be used for growing fuel, fodder and fruit bearing trees (Fig. 9.12).

It is obvious from table 9.3 that 14.7 per cent area of village Nalni lies in the medium quality (B_2) land, 62.05 and 23.25 per cent area is respectively occupied by poor quality C_1 and C_2 land. B_2 land is utilized for wheat and barley in rabi and potato and peas in kharif seasons. Some portion of the village is used for growing orchards. C_1 quality land is partly left as fallow and partly covered by oak and deodar trees. C_2 land is used for grazing in summers and growing grasses in rainly season. Keeping the environmental conditions in mind the village land is suitable for developing fruit culture especially for apple and pears (Fig. 9.15B).

Thus, the proportion of good quality land is higher in dun floor village as compared to mid slope category village while the upland village has not a single patch of good quality land.

9.4 LANDUSE CHANGES

Landuse figures for 1954-55 are obtained from land record offices of Nainital, Kaladhungi and Ramnagar and those for 1986-87 are based on field works. Table 9.4 shows that the area of unproductive uses—settlements, roads, paths, waterbodies, etc. increased by 0.69 per cent in the dun floor but it shows a decreasing trend in mid and upland villages by 2.22 and 0.11 percents respectively. Because of the unavailability of modern facilities in upland and mid slope villages, the inhabitants use to migrate to the dun floor or other developed centres in search of jobs and better facilities. The house holds migrated to nearby dun floor, leave their houses and lands as a waste. The culturable waste land is being brought under cultivation in the dun floor while it is increasing in mid slopes. The cultivable waste area decreased by 53.02 per cent in the upland village, infact it is due to inclusion of forest area in 1955-56 by the Revenue Department. Based on field work the figure for 1985-86 excludes the forest area otherwise waste land would have shown an increasing trend in this village also. It is a striking point that the area of orchards increased by 2.69 per cent in uplands, 1.17 per cent in mid slope and 0.85% in dun floor. The land in dun floor is suitable for cultivation while upland and mid slope land is being devoted to the orchards. The cultivated area increased in dun floor while it shows a decreasing trend in upland and mid slope villages (Table 9.4).

Table 9.4: Changing Landuse in the Sample Villages (1954-55 to 1986-87)

Landuse Class	*Sample Villages*		
	Dun Floor (Patkot)	*Mid Slope (Amgarhi)*	*Uplands (Sigarhi)*
Forest area	. .	. .	+53.60
Area not available for cultivation	+0.69	–2.22	–0.11
Cultivated waste area	–18.68	+2.63	–53.02
Orchards	+.85	+1.17	+2.69
Cultivated area	+17.14	–1.58	–3.16

9.5 LANDUSE

The percentage of area under forest is 13.21 in dun floor, 9.34 per cent in mid slope and 53.60 per cent in upland villages (Table 9.5). The proportion or area under forest is highest in upland, while lowest is in midslope village because in Kotadun this zone is very hazardous and the growth of vegetative cover is very nominal. The percentage of cultivated land is highest in dunfloor village, followed by a small proportion in mid slope village and only 12.48 per cent area is under cultivation in upland village. Thus, the area under agriculture decreases from dunfloor to upland villages. The topographic as well as environmental conditions of dunfloor are very favourable for growing any type of crops while the conditions of midslope and upland areas are not so favourable. The percentage of area under not available for cultivation is 4.12 in dunfloor, 7.70 in mid-slope and 2.83 per cent in upland villages. The proportion of cultivable waste is lowest in dunfloor (13.20%) while 45.63 per cent in mid slope and 28.40 per cent in upland village. Highest percentage of waste land is in mid-slope category village. It is most problem oriented area with low proportion of forest cover, and consequent soil erosion, landslides, masswasting, and migration. Owing to these causes there is a tendency of cultivating waste lands. The proportion under orchards and fruits is the lowest in dunfloor and highest in upland village.

The climatic conditions are very favourable to grow tropical and temperate fruits in the region (Fig. 9.1, 9.2, 9.7, 9.8, 9.13 A and B).

Table 9.5: Area Under Different Landuse Classes in the Sample Villages (1986-87)

Landuse Class	*Sample Villages*					
	Dun Floor (Patkot)		*Mid Slope (Amgarhi)*		*Uplands (Sigarhi)*	
	(ac)	*(%)*	*(ac)*	*(%)*	*(ac)*	*(%)*
Civil Forest	118.43	13.21	24.32	9.34	130.94	53.60
Area not available for cultivation	36.92	4.12	20.05	7.70	6.92	2.83
Culturable waste area	118.41	13.20	118.86	45.63	69.38	28.40
Orchards	8.58	0.96	3.69	1.42	6.56	2.69
Cultivated area	614.04	68.51	93.58	25.91	30.50	12.48
Total	896.38	100.00	260.50	100.00	244.30	100.00

9.6 CROPPING PATTERN

The cropping pattern of the sample villages is given in table 9.6. It is seen that 95.15 per cent area is cropped during kharif season in dunfloor while 98.58 and 94.27 per cent area is under kharif crops in mid slope and upland villages. Remaining 4.85, 1.42 and 5.77 per cent area is respectively left as fallow in these villages. The table also reveals that 92.45, 95.40 and 31.90 per cent of the cultivated area is given to grain crops while other crops—vegetables, ginger, spices, potato etc., occupy 2.70, 3.18 and 62.33 per cent area. It is notable that the environmental conditions are not permissible to grow kharif grain crops in the upland village therefore, maximum area (31.90%) is devoted to peas and corn. Paddy and soyabean are the dominant crops in dunfloor and mid-slope villages occupying 55.46, 28.04 and 78.41, 16.99 per cent area. Paddy includes basmati, binduli, anjana, china 4, tiachuing, indrasan etc. Basmati is luxurient and cash crop in dunfloor. Other varieties of paddy cover largest portion because these make a major proportion of the villager's diet and also serve in fatching them cash. Paddy is sown in three methods—ropai, saya and broadcast, Madua-urd mixed cropping pattern occupying only 7.90 per cent of the cultivated area is seen in dunfloor. Madua, bhatt, gahat, urd, soyabean are common crops and grown on all altitudes in the kharif season.

It is also observed that madua and bhatt occupy highest proportion in mid-slope village. It is followed by unirrigated paddy (*Chaity dhan*). Paddy and bhatt both are hardy crops and are grown in draught conditions also. Agronomically bhatt is very beneficial crop because being leguminous crop it adds nitrogen to the soil. Bhatt is used as pulse in the human diet. It is seen that in the kharif season 56.51 and 73.38 per cent of the cultivated area is under substantial crops respectively in dunfloor and midslope villages, remaining 28.04 and 16.99 per cent remains under commercial crop of soyabean while 31.90 per cent of the cultivated area is devoted to commercial crops of peas and corn in the upland village. Corn is also used to feed livestock as a concentrate while it is widely sold in summers in the tourist places of Nainital. Almost other crops viz. vegetables, spices are grown in small patches mainly near the settlements where proper facilities of irrigation and weeding are available.

A statement of village wise area occupied by various uses in rabi season is also shown in table 9.6. The rabi net cropped area is less than that in kharif season. It is only due to the scarcity of water for irrigation and unavailability of labour. Net 93.65, 95.45 and 90.82 per cent area is cropped and remaining 6.35, 4.54 and 9.18 per cent area is left as fallow respectively in dunfloor, midslope and upland villages. Wheat local, HYV and mixed with lahi or lentil, are common crops in dun, mid and top villages. Out of total cultivated area 70.30, 75.05 and 79.54 per cent area is given to wheat crop in dunfloor, mid slope and upland villages. Barley is grown in midslope (4.45%) and upland (6.56%) villages in small portion. Barley is staple diet. Barley was a foodgrain of poor family. It is also fed to livestock as a concentrate. Barley and wheat (local) both are hardly crops and not only grown in any elevation but

Table 9.6: Existing Cropping Pattern in the Sample Villages (1986-87)

Crops	*Sample Villages*					
	Dun Floor (Patkot)		*Mid Slope (Amgarhi)*		*Uplands (Sigarhi)*	
	(ac)	*(%)*	*(ac)*	*(%)*	*(ac)*	*(%)*
Kharif Season						
Grain Crops	567.69	92.45	89.28	95.40	9.73	31.90
Paddy	340.53	55.46	73.38	78.41	..	..
Soyabean	172.21	28.04	15.90	16.99	..	..
Madua-urd	48.51	7.90	..	..	..	..
Peas	..	..	..	..	6.77	22.20
Pulses	6.44	1.05	..	..	..	..
Corn	..	..	..	..	2.96	9.70
Others	16.63	2.70	2.97	3.18	19.01	62.33
Other Vegetables	16.63	2.70	2.97	3.18	1.01	3.31
Potato	..	..	..	..	18.00	59.02
Fallow	29.72	4.85	1.33	1.42	1.76	5.77
Total	614.04	100.00	93.58	100.00	30.50	100.00
Rabi Season						
Grain Crops	567.04	92.34	87.04	93.01	26.26	86.10
Wheat	431.65	70.30	70.23	75.05	24.26	79.54
Wheat-lahi	78.55	12.79	..	..	..	..
Lahi	44.90	7.31	1.38	1.47	..	..
Gram	9.19	1.50	..	..	..	..
Lentil	2.75	0.44	..	..	..	..
Wheat-lentil	..	..	10.12	10.81	..	..
Barley	..	..	4.16	4.45	2.00	6.56
Others	8.03	1.31	2.29	2.45	1.44	4.72
Vegetables	8.03	1.31	2.29	2.45	1.44	4.72
Fallow	38.97	6.35	4.25	4.54	2.80	9.18
Total	614.04	100.00	93.58	100.00	30.50	100.00

also seldom fail despite draught conditions. The percentage area under rabi vegetables viz. potato, onion, cauliflower, cabbage, peas, radish, tomato, brinjal, spices, garlic, tobacco etc. is 1.31 in dunfloor, 2.45 in mid slope and 4.72 in upland villages. Vegetables are cultivated in kitchen gardens because of greater labour requirements particularly for wedding and irrigation. Mostly vegetables are cultivated in good quality land. Like is kharif mixed cropping practised in this season also, but it is not so wide spread. Wheat is mixed with lahi as an extra supplier but not as an insurance against the failure of one crop. Mixed cropping is practised to built up the soil fertility in the dunfloor and midslope villages (Fig. 9.3, 9.4, 9.9, 9.10, 9.14A and B).

9.7 DOUBLE CROPPED AREA

The scarcity of cultivated land and pressure of population, on account of the exceeding birth and immigration in dun floor village have led the people to crop the land twice a year. The area cropped twice in rabi and kharif is known as double cropped area. As the table shows, 85.81, 94.04 and 85.05 per cent of total cultivated area is respectively cultivated twice a year in the dunfloor, midslope and upland villages (Fig. 9.5, 9.11 and 9.15A).

Table 9.7: Land-Man Ratio in the Sample Villages (1986-87) ***(area in acres)***

Particulars	*Sample Villages*					
	Dun Floor (Patkot)		*Mid Slope (Amgarhi)*		*Uplands (Sigarhi)*	
	Area	*Perca-pita*	*Area*	*Perca-pita*	*Area*	*Perca-pita*
Total area	896.38	0.80	260.50	0.80	244.30	2.07
Cultivated area	614.04	0.54	93.58	0.29	30.50	0.26
Net kharif cropped area	584.32	0.52	92.25	0.28	28.74	0.24
Net rabi cropped area	575.07	0.51	89.33	0.27	27.70	0.23
Gross cropped area	1159.39	1.03	181.58	0.55	56.44	0.47
Doubled cropped area	545.25	0.48	88.00	0.27	25.94	0.22

9.8 LAND-MAN RATIO

An attempt is made in table 9.7 to shows per head share of population in the various uses of the sample villages. It also reveals the burden of population on land in dun, mid and upland areas.

The per capita land in the total area of the dun floor, mid slope and upland villages is respectively 0.80, 0.80 and 2.07 acres. Maximum per capita share of 2.07 acres goes to upland village where population is nominal and village is spread over a large proportion almost bounded by reserved forests. The per capita land for cultivation is 0.54, 0.29 and 0.26 acres in dun, mid and upland villages. The per capita cultivated land is more in dun floor as compared with that in mid and upland villages. The per capita cultivated land decreases from dunfloor to uplands. Because the maximum area of dun village has been brought under cultivation owing to favourable agronomic facilities and level topography. Due to fallowing practices the per capita share is further reduced in the kharif and rabi seasons to 0.52 and 0.51 ac in dun, 0.28 and 0.27 ac in mid and 0.24 and 0.23 ac in upland villages. The per capita double cropped land in the dun, mid and upland villages is respectively 0.48, 0.27 and 0.22 acres. The per capita land in gross cropped land that is the amount of cultivated land supporting one person in dun, mid and upland villages is 01.03, 0.55 and 0.47 acres respectively (Table 9.7).

9.9 POPULATION PROFILES

Out of the total 28.31, 61.55 and 37.29 per cent population is engaged in primary rural occupation in the dun, mid and upland villages while 4.88, 2.15 and 4.23 per cent persons are working in allied services respectively. Remaining population is non-worker (children and old persons), but these persons are also dependant on agricultural workers. Those persons working in allied services, reside in other parts of the district or the country. They sent a little amount of money to their family members who spent it for payment of land revenue and buying some agricultural inputs like HYV seeds and fertilizers. However, the analysis of occupational figures reveal that the burden of population in dun and mid slope villages fall on agricultural land while upland villagers partly depend on agriculture and partly on other wages. The number of literate persons is more in dun and mid villages. Maximum of 89.83% persons belong to scheduled caste in upland village but it is observed in the Kotadun that they (Scheduled Caste) do not perform their hereditary occupation.

Table 9.8: Population Profile (1986-87)

Category	*Sample Villages*								
	Dun Floor (Patkot)			*Mid Slope (Amgarhi)*			*Uplands (Sigarhi)*		
	M	F	T	M	F	T	M	F	T
Total	536	591	1127	183	143	326	67	51	118
Literacy	251	231	482	107	45	152	32	8	40
Illiteracy	285	360	645	76	98	174	35	43	78
Workers	225	149	374	113	95	208	26	23	49
Non Workers	311	442	753	70	48	118	41	28	69
Agricultural Workers	170	149	319	106	95	201	21	23	44
Other workers	55	..	55	7	0	7	5	..	5
Scheduled Caste	140	120	260	40	36	76	60	40	106
No. of families	..	..	197	..	..	47	..	..	22
No. of house	..	..	190	..	..	41	..	..	18

T Total

M Male

F Female

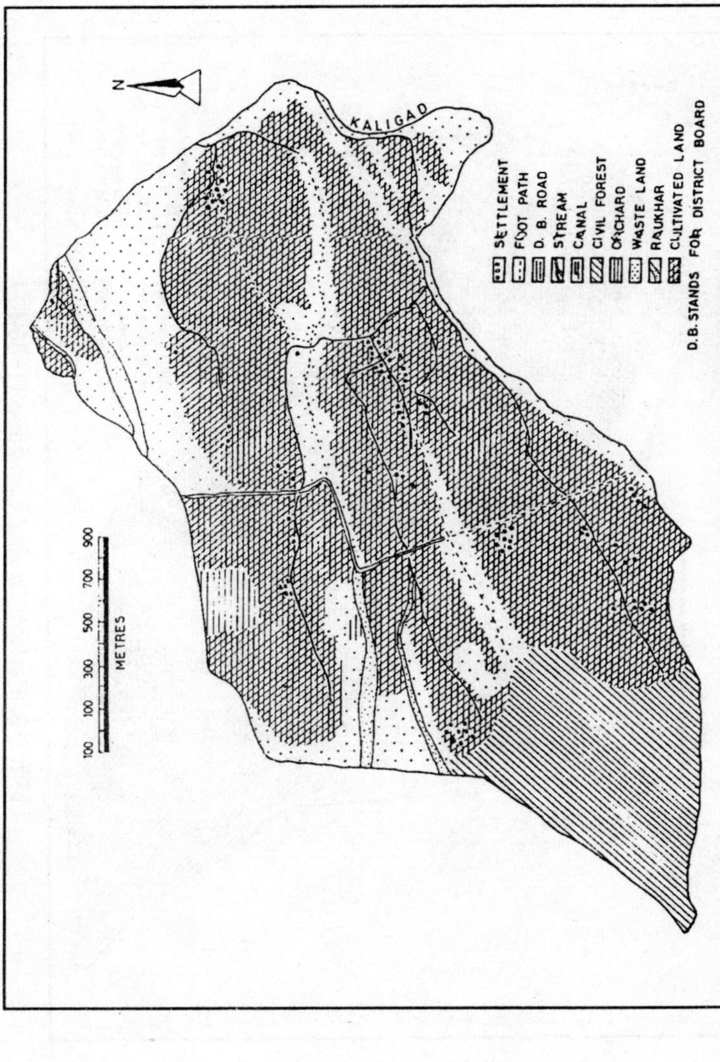

Fig. 9.1: Patkot: Major Landuse (1954-55)

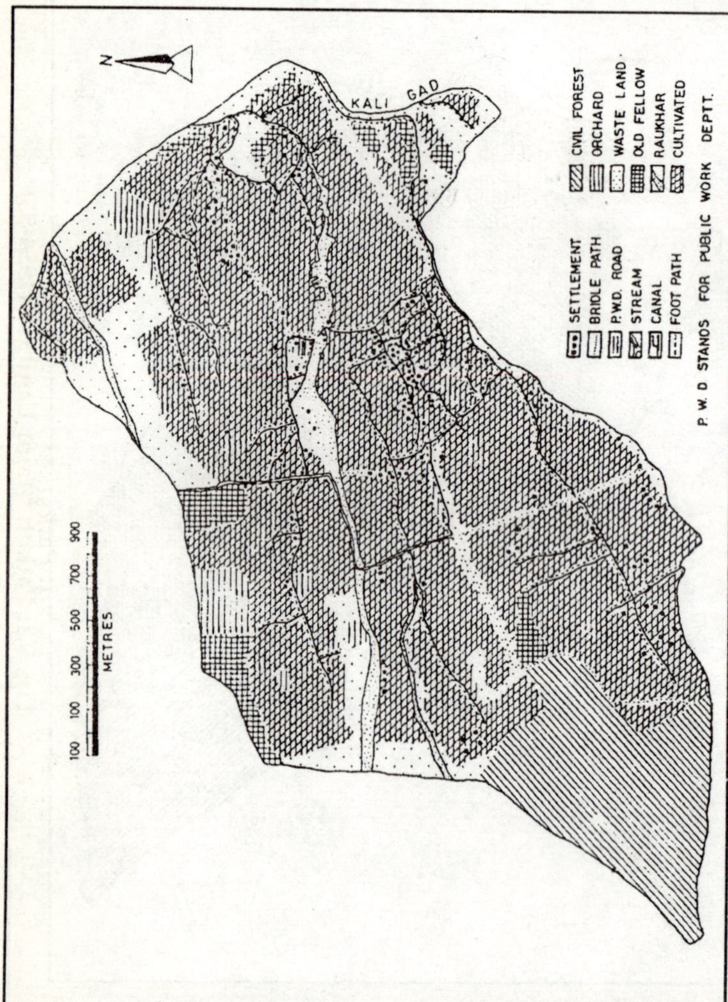

Fig. 9.2: Patkot: Major Landuse (1986-87)

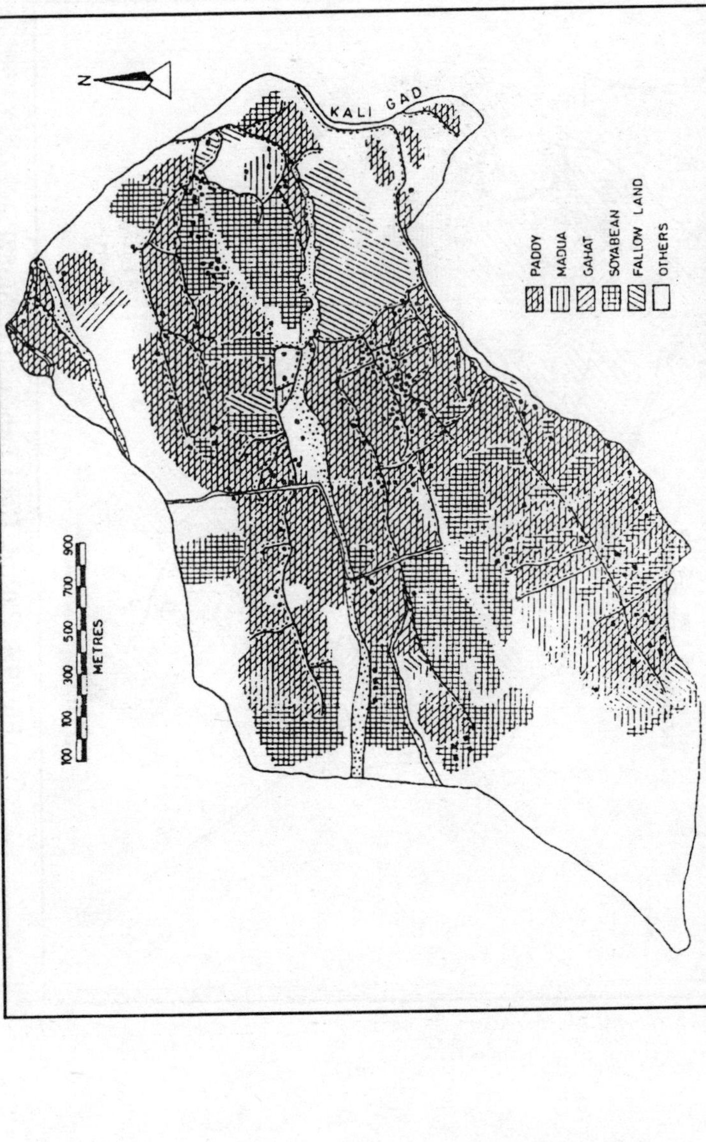

Fig. 9.3: Patkot: Kharif Landuse (1986)

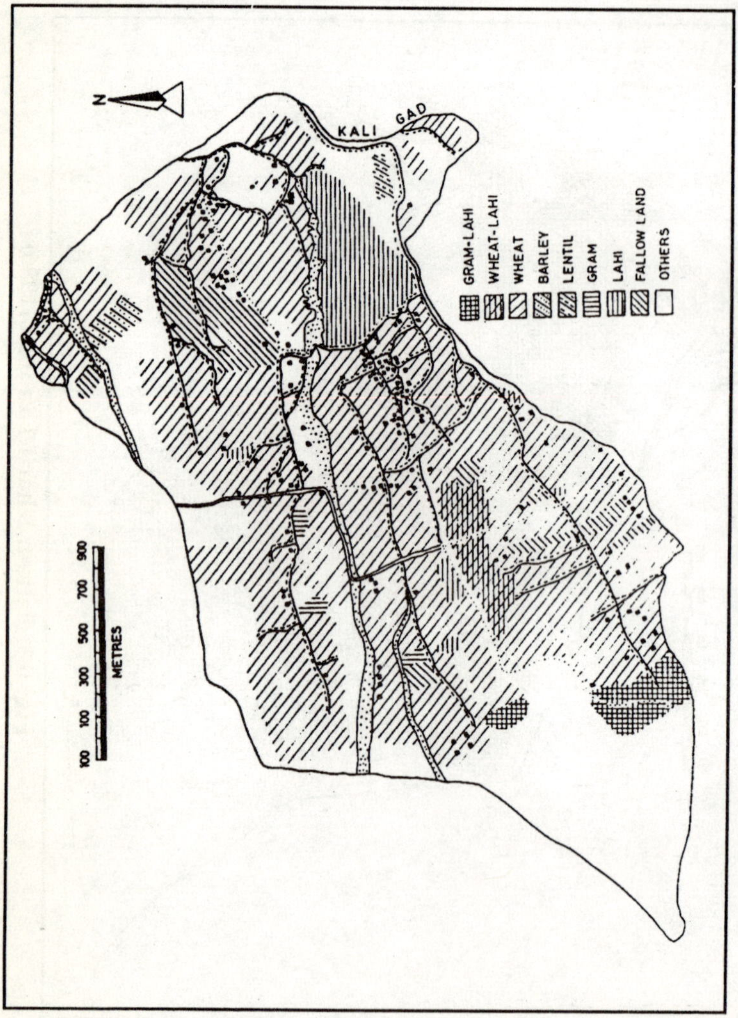

Fig. 9.4: Patkot: Rabi Landuse (1986-87)

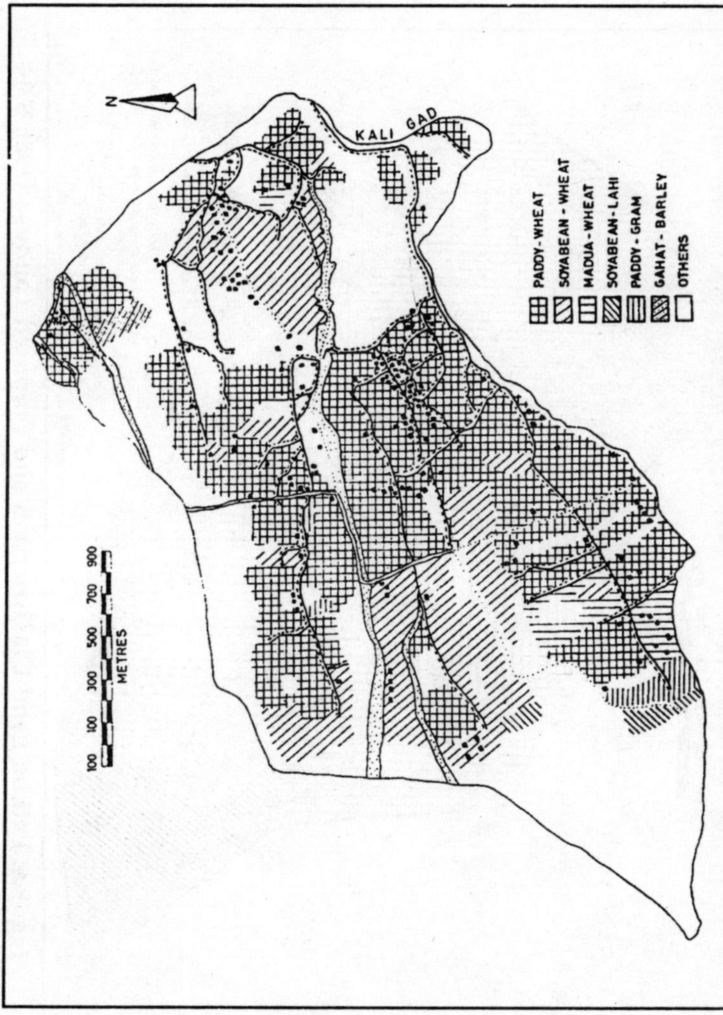

Fig. 9.5: Patkot: Double Cropped Area (1986-87)

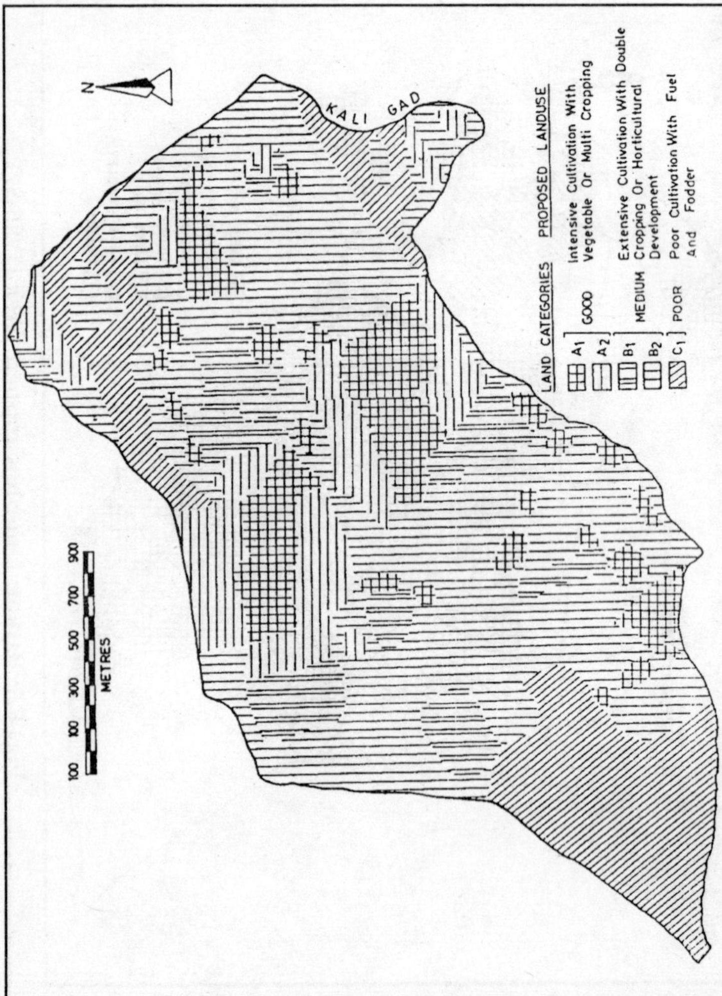

Fig. 9.6: Patkot: Land Classification and Proposed Landuse (1986-87)

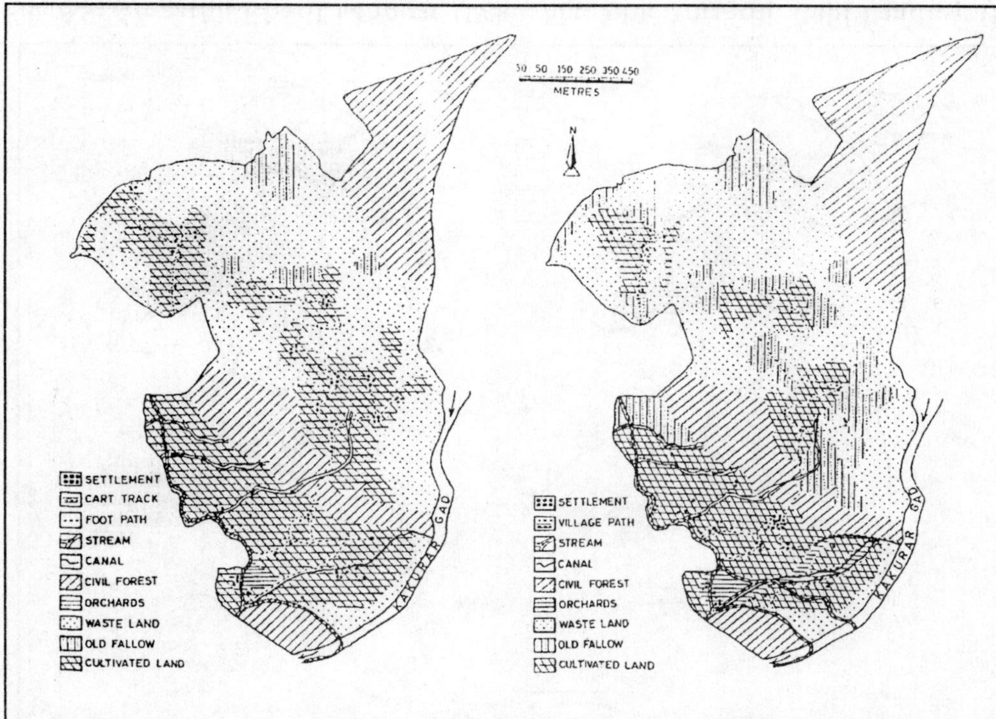

Fig. 9.7: Amgarhi: Major Landuse (1955-56). Fig. 9.8: Amgarhi: Major Landuse (1986-87)

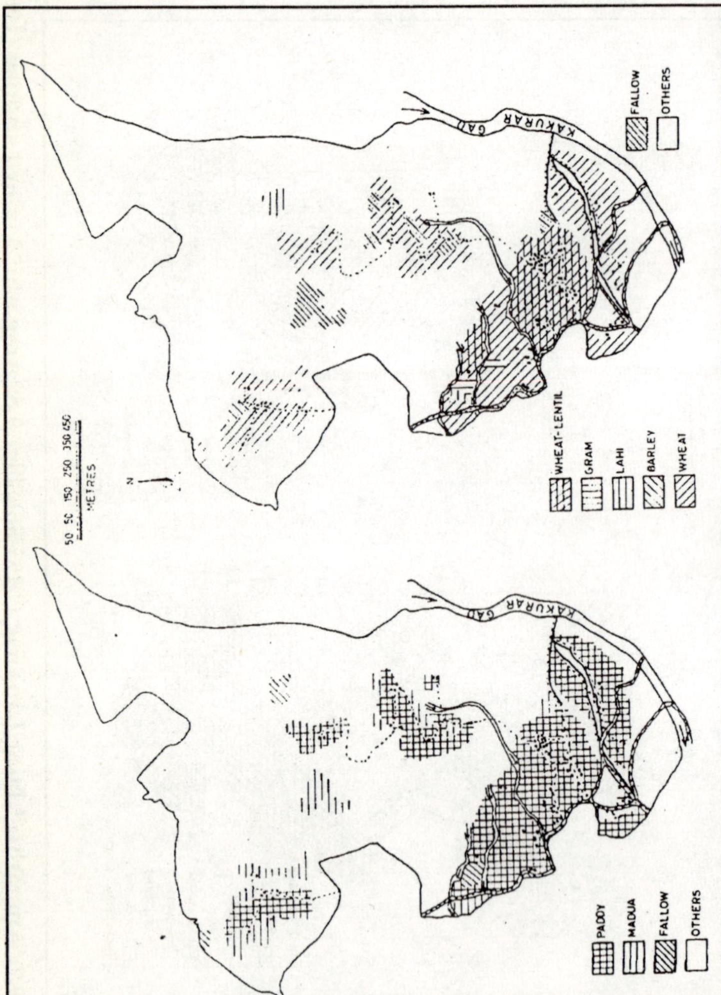

Fig. 9.9: Amgarhi: Kharif Landuse (1986). Fig. 9.10: Amgarhi: Rabi Landuse (1986-87)

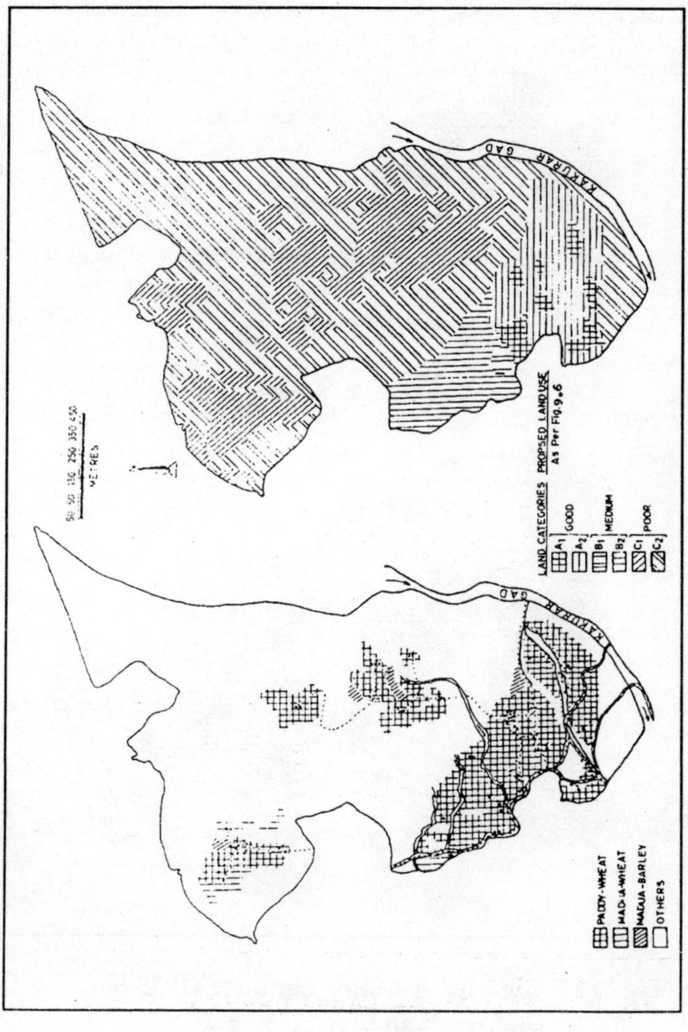

Fig. 9.11: Amgarhi: Double Cropped Area.

Fig. 9.12: Amgarhi: Land Classification and Proposed Landuse (1986-87)

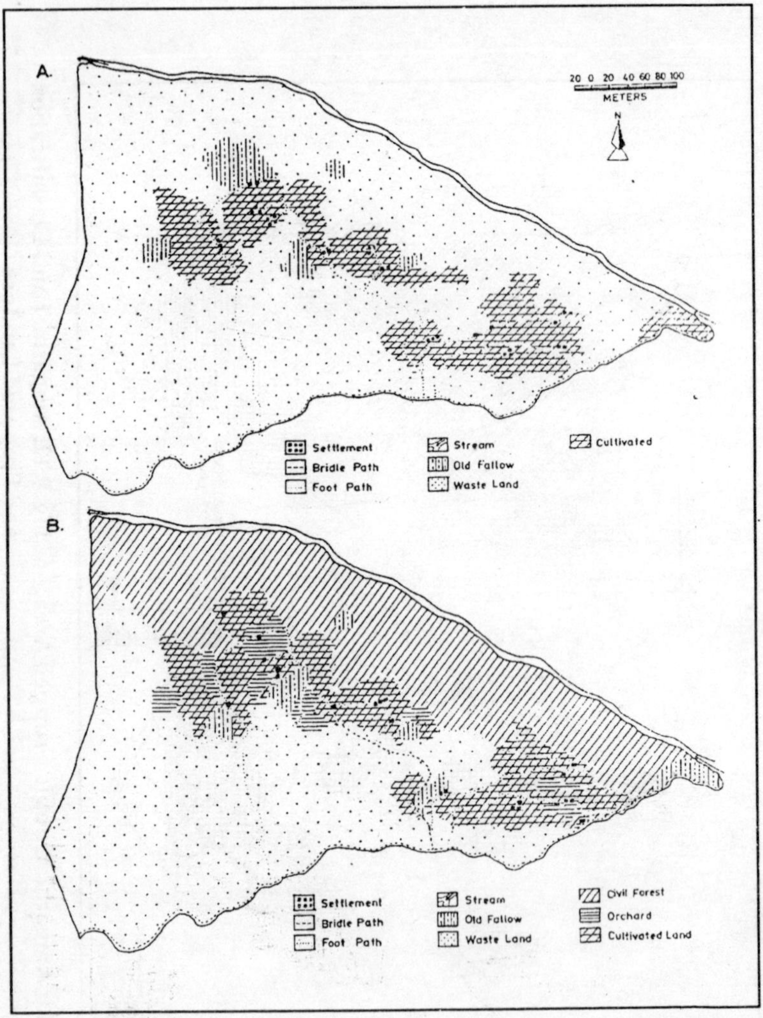

**Fig. 9.13: Sigarhi: A-Major Landuse (1955-56)
B-Major Landuse (1986-87)**

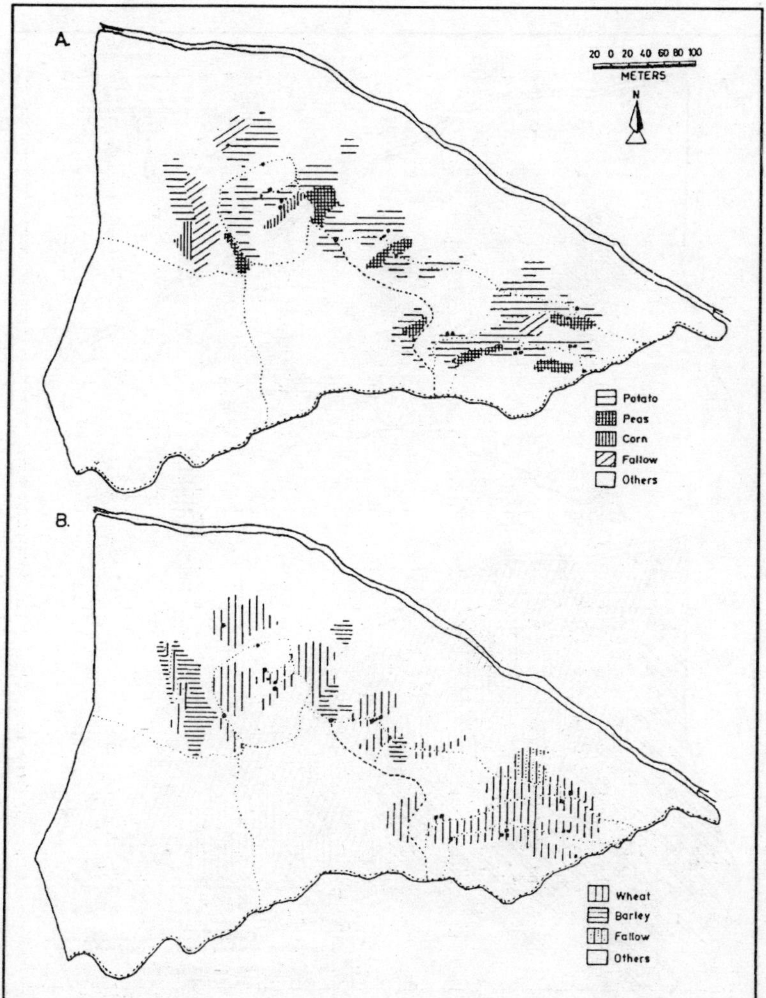

**Fig. 9.14: Sigarhi: A-Kharif Landuse (1987)
B-Rabi Landuse (1986-87)**

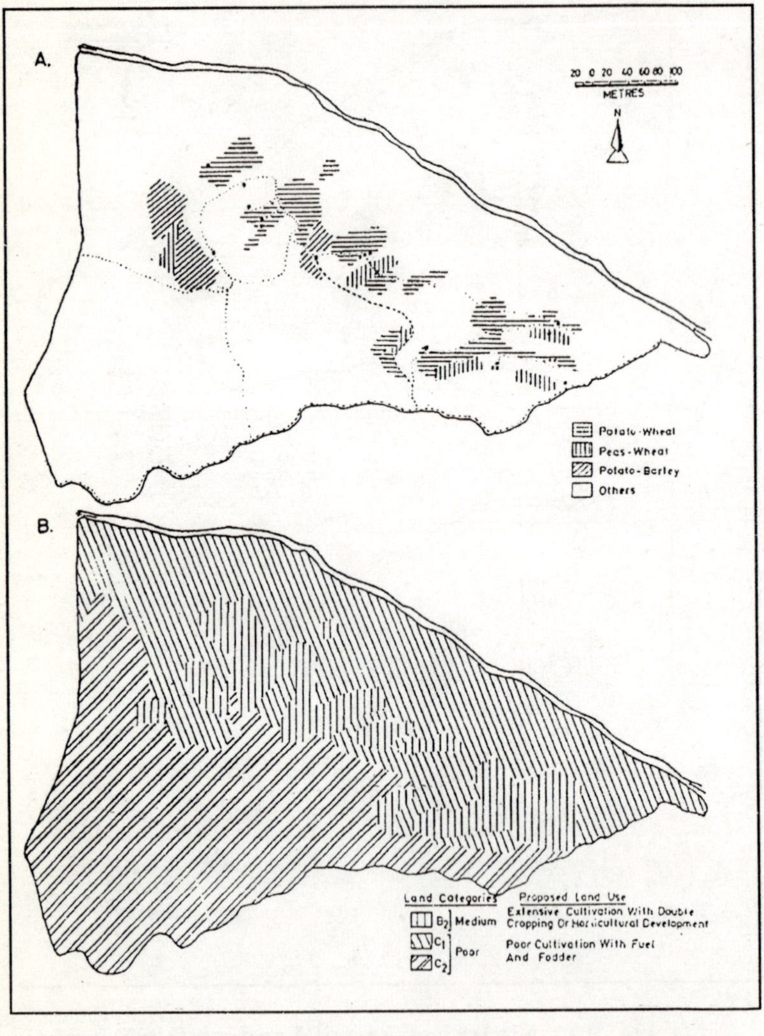

**Fig. 9.15: Sigarhi: A-Double Cropped Area (1986-87)
B-Land Classification and Proposed Landuse (1986-87)**

Chapter 10

Land Evaluation on the Basis of Land Capability Classification

The population is increasing not only in the sensitive, fragile and lofty Himalaya but also in the country and the world as a whole. Keeping pace with this serious and threatening problem, social scientists, planners, geographers, geologists, ecologists, environmentalists etc. are engaged in preparing and presenting various approaches in different parts of the world. But a plan which is suitable for one region, may not be suitable for the other. In this case, it is necessary for a planner that he should start the study of all the environmental constraints such as physico-cultural, socio-economic and traditions of the society of a particular region and than prepare development plans for that region. Geographers are emphasizing on watershed management approach for integrated development of mountainous regions. Several studies have established that the micro-watersheds are the appropriate planning units in mountainous regions. For detailed study some typical villages representing the characteristics are selected from the watershed. The Himalayan micro-watersheds possess almost similar conditions.

Keeping in view the above facets three villages are selected for detailed landuse study of Kotadun. The changing landuse (1954-55 to 1986-87), cropping pattern (1986-87) and land-man ratio is given in I and II technical reports and the comparative analysis of these aspects is included in Chapter 9. Present attempt evaluates the landuse of the sample villages interms of assessing what type of landuse exists in three different land categories and how it is influenced by environmental factors. It is based on land capability classification (Pant, 1987, Pant *et al.,* 1988). The maps pertaining to landuse and seasonal cropping patterns are superimposed turn by turn on land capability classification map and three main category uses—good (B), medium (B), and poor(C) are calculated and tabulated in tables 10.1 to 10.9.

On the basis of land evaluation one can quantify the faulty landuse in the village in particular and the region in general. Scientific and rational planning may be prepared on the basis of these findings.

10.1 LAND CAPABILITY AND LANDUSE

The use of land in Patkot is categorised into five groups forest, area not available for cultivation, orchards, culturable waste and cultivated area. Table 10.1 reveals that 13.21 per cent area is under civil forest lying in B category land. The forest area to some extent is degraded by the villagers for cultivation or for meeting the twin demand of fuel and fodder. Livestock population depends on civil and reserved forests for grazing, brauzing and lopping too. Land not available for cultivation covers an area of 4.12 per cent out of this 39.17, 54.39 and 6.45 per cent area respectively lies in A, B and C category lands. Although among these uses cultural water bodies, roads, settlements are the favourable determinants of the land quality.

A considerable area of 13.21 per cent lies under culturable waste land out of which 97.62 per cent area is under B category land. Old fallow land can easily be reclaimed and used for agricultural purposes. The waste land may be reclaimed with certain difficulty, on adopting high level of technological inputs with the help of soil and agricultural scientists. Net 0.96 per cent area is under orchards out of which 93.71 per cent area is in B category land. Fruit bearing trees spread around the settlements and water bodies mainly in medium class land. It is observed that some nearly settled families have established good orchards in their lands.

The cultivated area covers 68.50 per cent out which 45.37, 47.87 and 6.82 per cent area lies in A, B and C category lands. It is noteworthy that among the three land categories maximum area of the A category land is devoted to rabi and kharif cultivation due to several favourable conditions mainly well irrigational provisions, level land, optimum climatic conditions, application of new agricultural techniques (tractor, HYV, insecticides, pesticides, etc.), while 6.82 per cent area is devoted to cultivation of the C category land which is cultivated only due to ever increasing burden of population and land scarcity. The study reveals that the per capita land for cultivation is 0.54 acres which reflects the scarcity of agricultural land.

On the other hand 32.97 per cent area is under A category land out of which maximum proportion of 94.26 per cent is devoted to cultivation while 4.89, 0.72 and 0.13 per cent area is respectively in not available for cultivation, culturable waste and orchards. It is very striking point that on the basis of its suitability and quality of the land is devoted to different uses according to its utility. Best category land is mainly devoted to best uses i.e. intensive cultivation in rabi and kharif seasons while roads, settlements, gule and canals are the complimentary constraints of this lands encompassing an area of 4.89 per cent of good quality land and culturable waste area is negligible (0.72%) in this category.

B category land possesses 61.99 per cent of the total area of the village out of which 21.31 per cent area is in civil forest. The condition of forest area is under deterioration. Due to unscientific exploitation particularly lopping, brauzing and grazing, these areas are used to get covered with unwanted bushes (lantana). Maximum area accounting 20.80 per cent the total B category land lies under culturable waste land. It can be reclaimed and used with some difficulty for the cultivation if heavy dose of manures and irrigational facility are provided and the slopes are converted into terraces. Keeping in view the fertility and productivity of the soil 1.45 per cent area is devoted to various fruits—mangoes, bananas, guavas, jackfruits, lemons pears, etc. Net 52.83 per cent area of the total medium quality land is under cultivation. The landuse figures of this land category incorporate about 50.00 per cent area in other uses—forests, area not available for cultivation, orchards etc. While rest 50.00 per cent area is under cultivation. It is due to the reasons of tinny field size, gentle slope, medium erosion intensity, medium fertility, small size of land holdings, mismanagement and carelessness of farmers etc.

Since several environmental condition exist there in the village for good and medium category lands only 5.04 per cent of the total area is under C category land, out of which 5.27 per cent is under the area not available for cultivation. Settlements, roads, paths generally around the good quality lands. Infertility and other restricted constraints lower down the quality of these lands. However, 92.87 per cent area of C class land is under cultivation. Small patches in different parts of the village are generally under this poor category land (5.04% of the total). Poor landuse pattern clearly indicates the subsistantial nature of agriculture. Few farmers are holding this category land and they are fulfilling all of their requirements from it. Although it is a faulty landuse pattern but it is the need of the cultivators and the society. This area is comprised of boulders and pebbles. About 0.38 per cent area is under gardens. Gardens are generally situated away from the settlements and thus receive less attention for ploughing, manuring and weeding.

Table 10.1 incorporates a small portion of the land under other category uses, there exist some prospects for development through using heavy dose of manures and increasing the irrigational facility. The demonstration programme are required to be launched by the environmentalists, ecologists, economists, geographers and soil scientists for improving the landuse.

Table 10.2 reveals that 9.34 per cent area of the Amgarhi village is under civil forest, though it is not covered by trees but the saplings of sal and others are emerging naturally. This area belongs to C category land. Area not available for cultivation covers an area of 7.7 per cent of the total area, out of which 13.82, 21.25 and 64.93 per cent area is respectively in A, B and C category lands. Maximum area of this land lies under natural water bodies—drains and stream beds because the weak lithology leads the seasonal drains to widen or change their coarses every year. A considerable area accounting 45.63 per cent of the village lies under culturable waste land, out of which 94.78 per cent area lies under C category land. Culturable waste land is sub-divided into two classes-permanent fallow and old fallow lands. Owing to the low fertility of soil and unavailability of labours, people leave the fields as fallow for regaining of soil fertility. It is also observed that young generations owning these lands have migrated to institutionally developed places possessing fertile soil. Remaining 5.0 and 0.22 per cent area is under B & C category lands. A and B category lands are cultivated in maximum limit. Those fields situated away from settlements are left as fallow during rabi season.

Orchards covering an area of 1.42 per cent of the total, lies under A category land while a number of fruit and fodder trees are not included in this account as these trees are grown on the dolls of the cultivated fields and along the drains. Mangoes, guavas, pears, bananas, lemons etc. are important fruits grown there. Economically well to do families fulfil their fruit requirements from outside supplies. Due to unavailability of marketing and processing facilities people are not aware of growing and developing orchards but some orchards are now in growing stage which are expected to give returns after some years. Some fruit trees are suffering from insects and plant diseases for which there is no any Garden Protection Unit.

About one third (35.91%) area of the village is under cultivation out of which 33.34, 37.98 and 28.68 per cent area lies under A, B and C category lands respectively. Owing to the low productivity of the soil and ever increasing pressure of population, cultivation exercises in all categories. The total production of the area do not fulfil the requirement of the villagers, they mostly depend upon exchange or market supplies of foodgrains from three months of post-harvest. Consequently people depend on other occupations—livestock rearing, lumbering, quarrying and part-time labours.

On the other hand, A category land covers 14.56 per cent of the total area out of which 82.28 per cent is under cultivation while 9.73, 7.30 and 0.69 per cent is under orchards, not available for cultivation and culturable waste respectively. Owing to its suitability or betterness in comparison to others, maximum area is brought under plough while waste land proportion is very nominal (0.69%), B category land encompasses an area of 17.56 per cent of total out of which 77.70, 12.99 and 9.31 per cent area is under cultivation, culturable waste and not available for cultivation respectively. It is very interesting to note that maximum (67.88%) area of the village lies under C category land. 13.75 per cent area is occupied by civil forest which is under the

Table 10.1: Patkot: Major Landuse in Different Land Categories (1986-87)

S. No.	Landuse	Total area		Land Categories					
		ac	%	A	B	C	A	B	C
				Percentage of Landuse class			Percentage of Land category		
1.	Forest area	118.43	13.21	–	100.00	–	–	21.31	–
2.	Area not available for cultivation	36.92	4.12	39.16	54.39	6.45	4.89	3.61	5.27
3.	Culturable waste area	118.41	13.21	1.82	97.62	0.56	0.72	20.80	1.48
4.	Orchards	8.58	0.96	4.31	93.71	1.98	0.13	1.45	0.38
5.	Cultivated area	614.04	68.50	45.37	47.81	6.82	94.26	52.83	92.87
	Total	896.38	100.00	32.97	61.99	5.04	100.00	100.00	100.00

Table 10.2 Amgarhi: Major Landuse in Different Land Categories (1986-87)

S. No.	Crop	Total area		Land Categories					
		ac	%	A	B	C	A	B	C
				Percentage of Landuse			Percentage of land category		
1.	Forest	24.32	9.34	–	–	100.00	–	–	13.75
2.	Area not available for cultivation	20.05	7.70	13.82	21.25	64.93	7.30	9.31	7.36
3.	Culturable waste land	118.86	45.63	0.22	5.00	94.78	0.69	12.99	63.71
4.	Orchards	3.69	1.42	100.00	–	–	9.73	–	–
5.	Cultivated	93.85	35.91	33.34	37.98	28.08	82.28	77.70	15.18
	Total	260.50	100.00	14.56	17.56	67.88	100.00	100.00	100.00

Table 10.3 Sigarhi: Major Landuse in Different Land Categories (1986-87)

S. No.	*Crop*	*Total area*		*Land Categories*			
		ac	%	*B*	C	*B*	C
				Percentage of landuse		*Percentage of land category*	
1.	Forest	130.94	53.60	–	100.00	–	62.89
2.	Area not available for cultivation	6.92	2.83	13.29	86.71	2.56	2.88
3.	Culturable waste	69.38	28.40	24.88	75.12	48.05	25.01
4.	Orchards	6.56	2.69	–	100.00	–	3.15
5.	Cultivated	30.50	12.48	68.10	41.84	49.39	61.20
	Total	244.30	100.00	14.70	85.30	100.00	100.00

Table 10.4: Patkot: Kharif Cropping Pattern in Different Land Categories (1986-87)

S. No.	Crop	Total area		Land Categories					
		ac	%	A	B	C	A	B	C
				Percentage of crop			Percentage of category		
1.	Paddy	340.53	55.46	54.35	43.81	1.89	66.43	50.83	14.91
2.	Soyabean	172.21	28.05	45.63	52.62	1.75	28.21	30.87	7.20
3.	Madua-Urd	48.51	7.90	–	81.65	16.60	0.31	13.49	19.20
4.	Pulses	6.44	1.05	–	29.19	70.81	–	0.64	10.87
5.	Other vegetables	16.63	2.71	63.92	22.55	13.53	3.82	1.28	5.37
6.	Fallow	29.72	4.83	11.51	28.60	59.89	1.23	2.89	42.45
	Total	614.04	100.00	45.37	47.81	6.82	100.00	100.00	100.00

Table 10.5: Amgarhi: Kharif Cropping Pattern is Different Land Categories (1986-87)

S. No.	*Crop*	*Total area*		*Land Categories*					
		ac	*%*	*A*	*B*	*C*	*A*	*B*	*C*
				Percentage of crop			*Percentage of land category*		
1.	Paddy	73.38	78.42	37.94	46.29	15.77	89.23	95.58	43.11
2.	Madua	15.90	16.99	2.45	9.87	87.67	1.25	4.52	51.94
3.	Vegetables	2.97	3.17	100.00	–	–	9.52	–	–
4.	Fallow	1.33	1.42	–	–	100.00	–	–	4.96
	Total	93.58	100.00	33.34	37.98	26.68	100.00	100.00	100.00

Table 10.6: Sigarhi: Kharif Cropping Pattern in Different Land Categories (1986-87)

S. No.	Crop	Total area		Land Categories			
		ac	%	B	C	B	C
				Percentage of crop		Percentage of land category	
1.	Crop	2.96	9.70	100.00	–	16.69	–
2.	Potato	18.00	59.02	55.56	44.44	56.37	62.70
3.	Peas	6.77	22.20	55.69	44.31	21.25	23.51
4.	Others	1.01	3.31	100.00	–	5.69	–
5.	Fallow	1.76	5.77	–	100.00	–	13.79
	Total	30.50	100.00	58.16	41.84	100.00	100.00

Table 10.7: Patkot: Rabi Copping Pattern in Different Land Categories (1986-87)

S. No.	Crop	Total area		Land Categories					
		ac	%	*A*	*B*	*C*	*A*	*B*	*C*
				Percentage of crop			*Percentage of category*		
1.	Wheat	431.65	70.30	53.78	45.77	0.45	83.33	67.32	4.60
2.	Wheat-lahi	78.55	12.78	39.92	51.92	8.16	11.26	13.89	15.29
3.	Lahi	44.90	7.31	4.32	95.68	–	0.70	14.63	–
4.	Gram	9.19	1.50	45.59	20.46	33.95	1.50	0.64	7.44
5.	Lentil	2.75	0.45	–	20.00	80.00	–	0.18	5.25
6.	Other vegetables	8.03	1.31	88.16	–	11.84	2.54	–	2.26
7.	Fallow	38.97	6.35	4.77	25.12	70.11	0.67	3.34	65.16
	Total	614.04	100.00	45.37	47.80	6.82	100.00	100.00	100.00

Table 10.8: Amgarhi: Rabi Cropping Pattern in Different Land Categories (1986-87)

S. No.	Crop	Total area		Land Categories					
		ac	%	A	B	C	A	B	C
				Percentage of crop			Percentage of land category		
1.	Wheat	70.23	75.05	44.43	37.05	18.52	100.00	73.21	48.47
2.	Wheat-lentil	10.12	10.81	–	14.43	85.57	–	4.11	32.27
3.	Barley	4.16	4.45	–	100.00	–	–	11.71	–
4.	Lahi	1.38	1.47	–	33.33	66.67	–	1.29	3.43
5.	Gram	1.15	1.23	–	100.00	–	–	3.24	-
6.	Vegetables	2.29	2.45	–	100.00	–	–	6.44	-
7.	Fallow	4.25	4.54	–	–	100.00	–	–	15.83
	Total	95.58	100.00	44.43	37.98	26.68	100.00	100.00	100.00

Table 10.9: Sigarhi: Rabi Cropping Pattern in Different Land Categories (1986-87)

S. No.	Crop	*Total area*		*Land Categories*			
		ac	%	*B*	*C*	*B*	*C*
				Percentage of crop		*Percentage of land category*	
1.	Wheat	24.26	79.54	52.56	71.87	47.44	90.20
2.	Barley	2.00	6.56	100.00	11.27	–	–
3.	Others	1.44	4.72	100.00	8.12	–	–
4.	Fallow	2.80	9.18	55.36	8.74	44.68	9.80
	Total	30.50	100.00	58.16	100.00	41.84	100.00

management of the villagers. It fulfils their fuel and fodder requirements. Net 7.36 per cent area lies under not available for cultivation out of which maximum proportion comes under natural water bodies (*nala, raukher etc.*). A considerable area of 63.71 per cent lies in waste land. This is due to several causes; firstly, educated persons employed in govt. and semi-govt. services, business, constructions etc. are migrated to other places and nobody wants to cultivate or purchase their lands left as waste, secondly, due to low fertility of the soil little output in ratio of input, inadequate irrigation and transport facilities, tectonic unstability and related soil erosion and mass wasting activities convert these lands as waste. It can be reclaimed with some difficulty. Horticultural activities for growing fruit, fodder and fuel bearing trees would be helpful in reclamation of these lands.

Only 15.18 per cent area under cultivation reveals the faulty landuse pattern in the village. The heavy pressure of population on limited land resources is also responsible for such faulty landuse pattern. The productivity of C class land is very poor. Chapter 9 incorporates the decreasing trend of cultivated land from 1955-56 to 1986-87.

It is obvious from table 10.3 that in the Sigashir there is 53.60 per cent area under dense oak and lofty deodar trees. The cause of this arboreal growth of forest is owing to its location away from human interference. Most of the villagers are residing in other nearby villages. Even then due to increasing population pressure, people clear the forest and other bushes for constructing field terraces and houses or kharak (Khatta). It is a process of settlement distribution in the hill area. Sigarhi village is settled in that process.

Area not available for cultivation is only 2.83 per cent of the total area out of which maximum area is covered with natural water bodies. It is very interesting to note that there are only B and C category lands while there is no single patch of A category lands, it is only due to climatic and topographic constraints which are absolutely disfavourable. Net 86.71 per cent area lies in C category land while 13.29 per cent area is under B category land of the total area not available for cultivation.

Culturable waste land possesses 28.40 per cent area out of which 75.12 per cent area lies in C and 24.88 per cent in B class lands. Waste and fallow land is used for grazing and growing grass (*Gajeo*) in rainy season. Maximum old fallow land lies in B class land while waste land (*banjar*) is in C class. Orchards particularly that of apples are grown in C category land. Cultivated land covers an area of 12.48 per cent of the total village, it indicates that people depend in other occupations and market supplies which are far from the village. They carry goods on their backs or on horse backs from the market and other villages. Out of total cultivated land 58.16 and 41.84 per cent land lies under B and C category lands.

On another hand, B category land covers an area of 14.70 per cent only while C category land occupies 85.30 per cent area of the total. Highest percentage (49.39%) of the cultivated area lies under B category while 48.05 per cent area is under unproductive uses (settlements and water bodies). Net 62.84 per cent area is covered by dense sal and deodar forest of the total C category land. Only 2.88 per cent area is devoted to paths and water bodies while 25.01 per cent area is left as waste. Only 6.12 per cent area is under cultivation in this category while 3.15 per cent area is under orchards.

10.2 LAND CAPABILITY AND KHARIF CROPPING PATTERN

It is evident from table 10.4 that 614.04 acres or 68.50 per cent area is under cultivation out of which 45.37, 47.81 and 6.82 per cent area lies in A, B and C quality lands in dun floor village of Patkot. The cultivated area is utilized for growing paddy, soyabean, madua, urd, and vegetables in kharif season. The area under these crops is 55.46, 28.05,

7.90, 1.05 and 2.71 per cent. Paddy and soyabean are the principal crops and cover 83.51 per cent of the total cultivated area. Remaining crops occupy nominal area. Paddy covers 54.35, 43.81 and 1.84 per cent area in A, B, and C category lands. Generally, paddy cultivation exercises in good and medium quality lands while coarse grains and inferior millets are grown in medium and poor quality lands. Inadequate infrastructural development, paucity of irrigational water, uncertainty of weather, ignorance of advance technology, high yielding variety seeds, insecticides and pesticides, lackness of post harvest technology for storage, processing and marketing etc. are responsible factors for low production, otherwise the physical environmental conditions are favourable for optimization of agricultural production.

Maximum area of vegetables lies in A category land because these are grown near settlements. But onion and gaderi are grown near the canals. A small proportion of cultivated area (4.83%) remains fallow in this season which is generally in C category lands.

Various varieties of paddy are grown in 66.43 per cent area of A category land while 28.21 per cent area is devoted to madua-urd and vegetables. In the B category land 50.83, 30.87 and 13.40 per cent area is occupied by paddy, soyabean and madua-urd, while 14.91, 7.20, 19.20, 10.87 and 7.37 per cent area of C category land is given to paddy, soyabean, madua-urd, pulses and vegetables. Highest percentage (42.45%) of this category land is left as fallow.

Net 98.58 per cent area of the Amgarhi village is cropped in kharif season (Table 10.5). Remaining 1.42 per cent area is left as fallow. Paddy is main crop which occupies 78.42 per cent of the cultivated area. Despite the several unfavourable conditions it is grown in all categories viz. 37.94, 46.29 and 15.77 per cent area of the total paddy is cultivated in A, B and C category lands respectively. It is noteworthly that highest percentage (46.29%) of the paddy area lies in B category while the yield per hectare is low. It is only due to the fact that maximum area of the village is under this category land. Madua is second staple crop covering an area of 16.99 per cent of the total cultivated land. Occupying 3.17 per cent area vegetables are generally grown near water bodies and settlements.

A category land is devoted to paddy (89.23%) and vegetables (9.52%) while madua covers 1.25 per cent area of Amgarhi village. Net 37.98% area of the cultivated land is under B category land out of which 95.58 and 4.42 per cent area is under paddy and madua respectively. Though 28.68 per cent area lies under C category land out of which 43.11 per cent area is brought under paddy though the productivity is very low. Maximum area (51.94%) is devoted to madua. Generally madua fields are laid above and far from the settlements in unirrigated and declivity areas. Remaining 6.96 per cent area is left as fallow.

In upland village of Sigarhi 12.48 per cent area is under cultivation out of which 59.02 per cent area is occupied by potato and 9.70 per cent by corn (Table 10.6). Although 55.56 per cent and 44.44 per cent potato area is respectively under B and C category lands. Corn is grown entirely in medium class land. Peas cover an area of 22.20 per cent of the total cultivated land out of which 55.69 and 44.31 per cent area is under B and C category. Others viz. turmeric, tomato, ginger etc. are grown 3.31 per cent in B class land. Rest 5.77 per cent area is left as fallow in this season.

Out of total B category land of Sigarhi village 56.37 per cent area is occupied by potato while 21.25 per cent area under peas. Corn (16.69%) is grown in B category land. Remaining cultivated area of this category land is occupied by spices and others vegetables. Net 62.70 per cent of C category land is devoted to potato while 23.51 per cent of its is under peas. Poverty compelles the people for exercising cultivation in poor quality lands. Remaining 13.79 per cent area is left as fallow in this season.

10.3 LAND CAPABILITY AND RABI CROPPING PATTERN

Wheat is the principal crop in rabi season not only in Patkot but also in the region as a whole. It covers 70.30 per cent of the total cultivated area while small proportion of the area is devoted to wheat-lahi mixed (12.78%), lahi (7.31%), gram (1.50%), lentil (0.45%) and vegetables (1.31%) (Table 10.7). Net 6.35 per cent area is left as fallow Lahi is the only source for obtaining cooking oil. Gram and lentil are the main pulses grown in this season. Due to lower yields pulses occupy smaller proportions and their requirements are fulfilled from market supplies. Rest 6.35 per cent of the cultivated area is left as fallow which is more than that of kharif season mainly due to shortage of water, low soil fertility and labours unavailability.

It is clear from table 10.7 that wheat receives 83.33, 67.32 and 4.60 per cent area in A, B and C category lands respectively. Wheat mixed with lahi gets 11.26, 13.89 and 15.29 per cent area. Lahi grown as separate crop occupies 14.63% area of B class and 0.70 per cent of C class lands other crops cover little portion of these category lands. Maximum area (65.16%) is left as fallow in the C category lands while it is only 3.34 and 0.67 per cent in B & A category lands.

Net 83.33 per cent of the A category land is devoted to various varieties of wheat while wheat mixed with lahi covers only 11.26 per cent area. Remaining 0.70, 1.50 and 2.54 per cent area is devoted to lahi, gram and vegetables respectively. Only 0.67 per cent area remains as fallow. Out of B category lands 67.32 per cent area is devoted to wheat. Lahi stands second accounting 14.63 per cent wheat-lahi (13.89%) ranks third in this category. The proportion of fallow land increases to 3.34 per cent while gram and lentil respectively occupy 0.64 and 0.18 per cent area. Vegetables are not grown in this category land. Only 6.82 per cent of total cultivated area is under C category lands. Out of which 4.60 per cent area is devoted to wheat. Maximum area (65.16%) of it is left as fallow. Wheat-Lahi covering 25.22 per cent area stand second. Gram, lentil and vegetables occupying 7.44, 5.25 and 2.26 per cent area are other crops grown in this land.

The seasonal cropping pattern of Patkot clearly indicates the faulty landuse pattern. It is only due to the subsistantial and traditional nature of agriculture. The villagers do not want to change their age old methods of cultivation. They devote their lands according to their needs. Illiteracy and less knowledge of advanced farming techniques are also responsible for such faulty landuse.

Table 10.8 shows that 75.05% area is under wheat and 10.81 per cent under lentil (wheat-lentil mixed) in the mid slope category village. Net 4.45 per cent area is devoted to barley which is used as on animal feeder. Out of total cultivated area 1.47, 1.23 and 2.45 per cent is devoted to lahi, gram and vegetables respectively.

It is very interesting to note that 33.34 per cent of the total area lies in A category land which is completely devoted to wheat while 73.21 per cent area of the B class land is also given to wheat. Barley occupies 11.71 per cent area of the B class land. Lahi occupies 1.29 per cent area, Gram and vegetables accounting 3.24 and 6.44 per cent area are also grown in medium class land. There is nominal cultivation of pulses and oil seeds. As people depend on market pulses, they do not fulfil their nutritional requirement and suffer from several deficiency diseases.

Table 10.8 reveals that 28.68 per cent area lies C class land, out of which wheat occupies 48.47 per cent and wheat-lentil 32.27 per cent. Lahi occupies 66.67% area, out of which 3.43 per cent is in C class land. Remaining 15.83 per cent of C category land is left as fallow.

The cropping pattern is faulty in several ways. Landuse reorganisation programme are required to be launched in order to educate the people about the capability and the best uses of the lands.

In uplands 79.54 per cent area is under wheat cultivation in rabi season out of which 52.56 per cent area respectively lies in B and C class lands in upland category village. Accounting 6.56 per cent area barley is grown in B category land. Vegetables accounting 4.72 per cent of the cultivated area, are also grown in B category land. Remaining 9.18 per cent area is left as fallow—55.36 per cent of B and 44.64 per cent of C category land (Table 10.9).

Out of total B category land 71.87 per cent is occupied by wheat while barley is grown in 11.27 per cent area. Remaining 8.12 and 8.74 per cent area is under vegetables and fallow land respectively. Out of total C class land 90.20 per cent area is devoted to wheat and 9.80 per cent area is left as fallow.

The cropping pattern of Sigarhi is also an indication of faulty landuse. The environmental conditions of the village are less favourable for growing seasonal crops but are appropriate for the development of horticulture and tourism.

Chapter—11
Agricultural Efficiency

Ever increasing population is the threatening problem, not only in the Kumaun but also in the Himalayan region as a whole, where more than 85 per cent population is engaged in agriculture. Thus, there is an urgent need to study the agricultural aspects of cropping pattern, carrying capacity, land-man ratio, productivity, nutritional status and agricultural efficiency. Present investigation is carried-on an experimental basis in Kotadun. The study highlights the major determinants responsible for low level of agricultural efficiency and suggests some measures for stepping it up. The efficiency is higher in dunfloor as compared to that in Lesser Himalayan units.

High pressure of population on the limited land particularly where there is no scope for expansion of cultivation, is very considerable issue for planners. To overcome this problem it is necessary that the agricultural efficiency should be stepped up in order to fulfil the needs of society including the nutritional requirements. The agricultural practices in Kotadun are rather traditional than scientific. In reference to human occupation people depend on agriculture, grazing, forestry and part-time quarring. Almost all category lands are under deterioration.

Agricultural efficiency is the best media to understand the regional imbalances of agricultural productivity which is determined by various factors such as physiographic, climatic, edaphic, hydrological, socio-economic, techno-organisational etc. in a particular area (Pant, 1988). The efficiency is in the sense that it obtains the maximum returns from the soil, the differentials in productivity per unit area being the result partly of natural advantage of soil and climate and partly of farming techniques (Shafi, 1960). It is reflected in yield per acres/hectares and total production from the particular unit (Sengupta and Sadasyak, 1961). Thus, it also referes to competence or capacity of land (Jalal, 1976). The term agricultural efficiency is relative one as the physical volume of out put per hectare in term or value or volume, is compared against the standarized value or volume (regional average) of farm produce is an areal unit (Pathak, 1977).

It is taken as a most significant index to show the spatial pattern of agricultural growth. As a matter of fact, the productivity per farm unit is different of any two geographical locations depending upon the different environmental conditions of each area. Owing to the dissimilarity in agricultural productivity in each unit results in differences in the efficiency of the unit, therefore, the need to develop a technique for the measuring of agricultural efficiency. To solve this hypothesis several techniques have been adopted by several social scientists since last thirties, such as (a) out put per unit area, (b) out put per unit of farm labour, (c) out put in relation to input and paid out coste in relation to out put, (d) standard nutrition needed and (e) potential production unit (PPU).

One of the earliest method is that of Kendall (1939) based on output per unit area by computing the ranking co-efficient. It was followed by Stamp (1958) and shafi (1960). The basic weakness in ranking co-efficient is that it takes into account the areal proportionate share of land under different crops within an areal unit only but the yield per unit area has not been taken into consideration. Spare and Despande (1964) attempted to remove this deficit of ranking co-efficient by giving weightage ranks to the area under different crops but the study also do not consider the per unit yield differences. Several other workers evolved methods for computing the agricultural efficiency. Most of these studies are confined to the regions of homogenous conditions like plain. Jalal (1976), Chand and Thakur (1985), Samant (1986), Pant (1988) and several other researchers also adopted these methods in Himalayan region. However, agricultural efficiency survey is not available for Kotadun. This part of Kumaun Himalaya is entirely different physico-cultural entity, distinguishable from other parts of the country.

METHODOLOGY

To evaluate the spatial variation in agricultural efficiency, the area has been divided into ten planning units with the help of slope, relief, and altitude maps. Data pertaining to area and crop yields are based on Revenue Records and field works respectively. Owing to certain limitations and availability of data, Bhatia's and Shafi's methods only are adopted the investigation although other techniques are equally useful for measuring the agricultural efficiency.

11.1 Bhatia's Efficiency Index

The principal eleven crops viz. paddy, soyabean, madua, jhungora, corn, kharif pulses, wheat, barley, gram, lentil and lahi are taken into consideration for computing the productivity index with the help of Bhatia's (1967) method. He made an attempt to determine the agricultural efficiency of Uttar Pradesh. He assumed that per hectare yields express all the physical and human factors connected with the production of crops and sharing of crops land among the various crops are reflecting the various factors expressed as following procedure:

$$Iya = \frac{Yc}{Yr} \times 100$$

Where, Iya is the yield index of crop 'a',

Yc is the hectare yield of same crop in the component area unit,

Yr is the hectare yield of same crop in the entire region, and

$$Ei = \frac{Iya \times Ca + Iyb \times Cb + \ldots\ldots\ldots + Iyn \times Cn}{Ca + Cb + Cc + \ldots\ldots\ldots + Cn}$$

Where, Ei is the agricultural efficiency index,

Iya, Iyb, Iyc +. . . . . . Iyn are the yield indices of various crops and Ca + Cb + Cc +. . . . . . Cn are the percentage of crop land under the different crops.

SPATIAL VARIATION IN AGRICULTURAL EFFICIENCY

The efficiency index for each unit has been computed the crop which are mapped figure 11.1 and tabulated in table 11.1. More than 90 per cent of gross cropped area is occupied by these crops. The indices very from a minimum efficiency index in U_9 (63.23) while the maximum in U_8 (155.57). Indices of agricultural efficiency are arranged into four groups, considering the index value for the region as 100.

Table 11.1 Degree of Agricultural Efficiency

S. No.	*Degree of efficiency*	*Index*	*Ranking of Units*
1.	Very Low	Below 70	U_9 (66.23), U_7 (66.42) U_5 (69.47)
2.	Low	70-100	U_3 (71.01), U_2 (75.34)
3.	Moderate	100-130	U_2 (112.84), U_6 (128.22)
4.	High	Above 130	U_4 (130.91), U_{20} (154.37) U_8 (155.57)

Table 11.1 incorporates that only five units viz, U_6 (128.22), U_2 (112.84), U_4 (130.91), U_{10} (154.37) and U_8 (155.57) are registered moderate to high agricultural efficiency. Because of suitable conditions viz. almost plain portion, availability of irrigation facilities and launching the new technologies for agricultural development while the diversified topographical conditions, less irrigational facilities, extreme climatic condition etc. in the U_1(75.34), U_3(71.01), U_5(69.47), U_7(66.42) and U_9(66.23) registered low the very low agricultural efficiency.

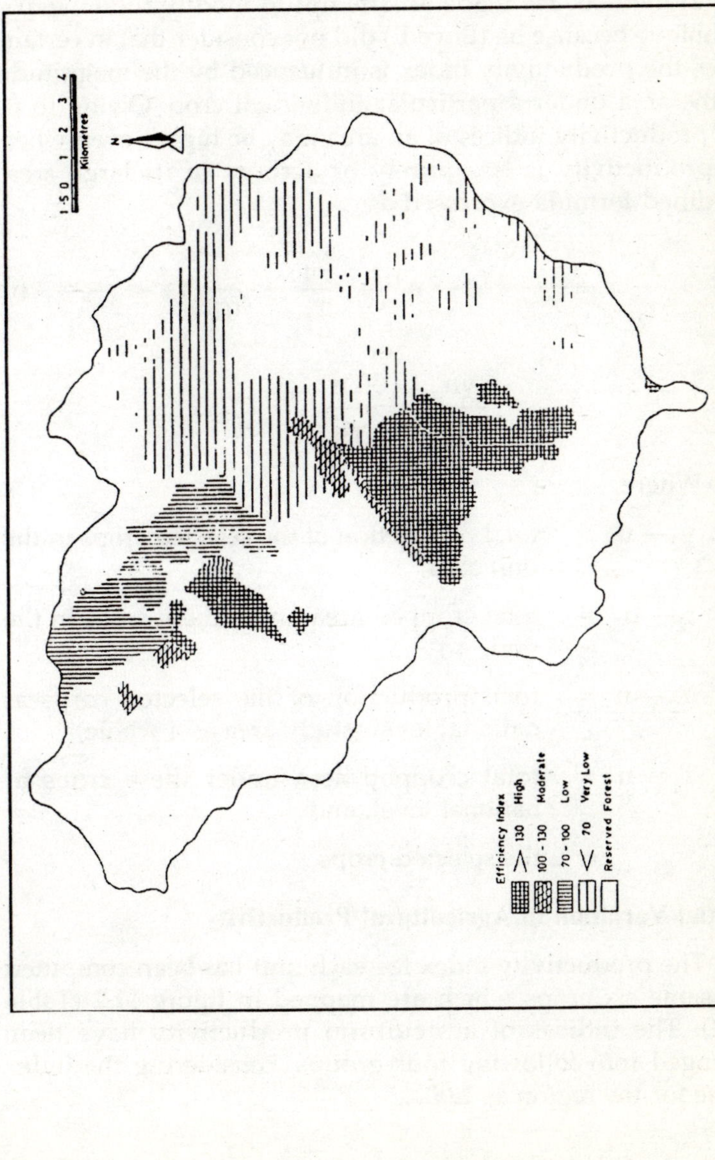

Fig. 11.1 Kotadun: Agricultural Efficiency (1985-86)

11.2. Shafi's Productivity Index

The principal six crops viz. paddy, wheat, madua, jhungora, barley and corn are taken into consideration for computing the productivity index with the help of Shafi's (1972) method. He made an attempt to modify the Enyedi's technique because he (Enyedi) did not consider that in certain cases the productivity index is influenced by the magnitude of the area under a particular individual crop. Owing to it, the productivity indices of an area may be higher even when its productivity is less purely or account of its large area. Modified formula expressed as:

$$\frac{y_1}{t_1} + \frac{y_2}{t_2} + \text{--------} n \;:\; \frac{Y_1}{T_1} + \frac{Y_2}{T_2} + \text{-------} n$$

or

$$\frac{yn}{tn} : \frac{Yn}{Tn}$$

Where,

$y_1, y_2 — n$ = total production of the selected crops in the unit area,

$t_1, t_2 — n$ = total cropped area under these crops in the unit,

$Y_1, Y_2 — n$ = total production of the selected corps at national level (study area as a whole)

$T_1, T_2 — n$ = total cropped area under these crops at national level, and

n = the selected crops.

Spatial Variation in Agricultural Productivity

The productivity index for each unit has been computed the same six crops which are mapped in figure 11.2 (Table 11.2). The indices of agricultural productivity have been arranged into following four groups, considering the index value for the region as 1.00.

(1) High productivity level, with index value more than 1.50,

(2) Moderate productivity level, with index value range between 1.00 to 1.50,

(3) Low productivity level, with index value range between 0.50 to 1.00, and

(4) Very low productivity level, with index less than 0.50.

Table 11.2 Unit-wise Productivity Index (1985-86)

Unit	*Productivity Index*
U_1	0.63
U_2	1.09
U_3	0.55
U_4	1.25
U_5	0.54
U_6	1.06
U_7	0.53
U_8	1.85
U_9	0.46
U_{10}	1.35

The spatial variation in productivity index is fairly large, ranging from the lowest for U_9 (0.46) to the highest for the U_8(1.85). It was also clear from the table and figure that five units U_8(1.85), U_{10}(1.38) U_4(1.25), U_2(1.09) and U_6(1.06) registered high productivity index. The low productivity units are U_1(0.63), U_3(0.55), U_5(0.54), U_7(0.53) and U_9(0.49).

11.3 Determinants of Agricultural Efficiency

Among the environmental factors terrain plays a significant role to determine the agricultural efficiency while cultural such as innovation of new agricultural techniques, high yielding variety seeds, chemical fertilizers, pesticides, insecticides and irrigational facilities also seem to be regulated by these natural environmental factors. It is noticed that below 800 m mean sea level and below 8° slope units show high agricultural efficiency while more than this limitation (800 m mean sea level and 8° slope) diverse topographic and extreme climatic conditions are responsible for low agricultural efficiency.

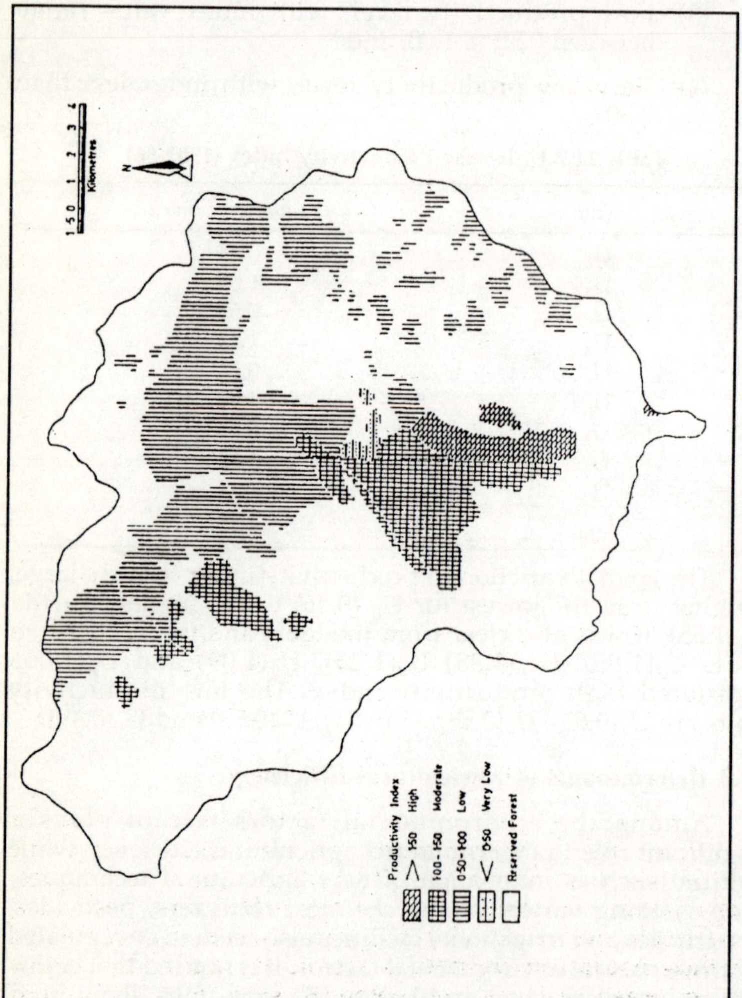

Fig. 11.2 Kotadun: Agricultural Productivity (1985-86)

There is subsistantial nature of agriculture where several natural and cultural factors governs the agricultural efficiency. Keeping in view the existing cropping pattern and prevailing facilities of the area some improvement is possible. Paddy and Wheat among food grains and soyabean among the cash crops should be taken as a primary crops in the high agricultural efficiency units (U_2, U_4, U_6, U_8 and U_{10}) and madua and barley should be grown as a secondary crops. In this proposed theoretical model maximum proportion of the areas recommended for wheat while relatively small area is suggested for paddy since this crop on the one hand can be grown in irrigated areas and on the other it requires more input in relation to wheat. The out put: input ratio for wheat and paddy are reported to be 3.26 and 0.59 respectively (Singh *et al.*, 1984). The status of the commercial crops ginger, peas, gram, lahi, potato which cover a negligible proportion should be taken as supplementary crops for the improvement of the present economic status of the people. Irrigational and technological facilities should be provided more and more in those units where agricultural efficiency is low.

Chapter—12

Food, Nutrition and Their Impact

Himalaya is under the grip of three major problems viz. soil erosion and mass wasting, dwindling water resource, and mismanagement of land resource. The problem of malnutrition and consequent deficiency diseases are not only due to the mismanagement or unscientific use of land resource, food preparation methods etc. but the ignorance about nutrition is also responsible to promote this fetal problem. Otherwise one can obtain adequate nutrition from locally available resources. To minimise this problem, it is inevitable to get sufficient knowledge of present situation of landuse, food availability, nutritional status and related deficiency diseases. On the basis of this study a suitable or feasible plan for improvement can be formulated. The ever increasing population on the limited resources is not a minor problem. Combined impact of all aforementioned problems accelerate the health hazard in the Himalaya is particular and the country in general.

Keeping in mind the above points a nutritional study of three sample villages is carried out in present attempt examining the general dietary habits, food systems, nutritional status and prevailing deficiency diseases.

Food Balance Sheets of the sample villages are prepared with the help of actual food production, import-export supplies. The food commodities are converted in term of energy and various other nutrients. Each of the nutrients is compared with the standard requirement suggested by the expert committee of Indian Council of Medical Research (ICMR) Hyderabad (Gopalan *et al.*, 1985). An attempt is made to correlate the deficiency of a specific food constituent with the prevailing deficiency diseases.

12.1 FOOD SYSTEMS

Dietary systems of Kotadun are similar to the whole Kumaun Himalaya. The people of the sample villages take their food in four times in a gastronomic day. Between the breakfast (*Kaleva*) in morning, and the tiffin in evening people take heavy lunch (*bhat*) because befor that they are engaged in strenuous agricultural or other activities as ploughing, manuring, lumbering, weeding, harvesting, threshing, collection of fuel and fodder, etc. Light meal like snaks are taken by few economically strong persons/families in late afternoon with tea. Other principal meal is taken in early night (dinner). Overwhelming population takes two meals only except during agricultural peak seasons. During the agricultural peak seasons, enjoying collected labours with the neighbours, some oily/fatty food is taken.

Generally the breakfast includes chapati or roti, seasonal vegetables and porridges (*suji, haluwa, kheer* etc.) in particular occassions. Chapaties are generally made of wheat flour, wheat and madua flour mixed (*lesuwa/lesa*) or grinded urd/gahat put inside the wheat floor (*beruwa*). The lunch consists of hand husked boiled rice (*bhat*) and pulse (*dal*). Among pulses gram, soyabean, lentil, pegion-pea, bhat (bleak soyabean), gahat, urd, malka etc. are generally consumed but poor families or for a taste, people take kari (*jholi*), grinded gahat (*dubke*), rice mixed with pulse (*khichari*), rice mixed with curd (*jaula*) or matha (*chchansia*) or mixed with grinded gahat (*rajara*), bhatt (*bhatti*), bari, mungari, churkani etc. Vegetables (*tapkiya*) in day time are occasionally or seasonally consumed. In evening, children take stale meal otherwise they take salt mixed bread either with tea, chatni, gur, ghee or vegetables. It is not taken

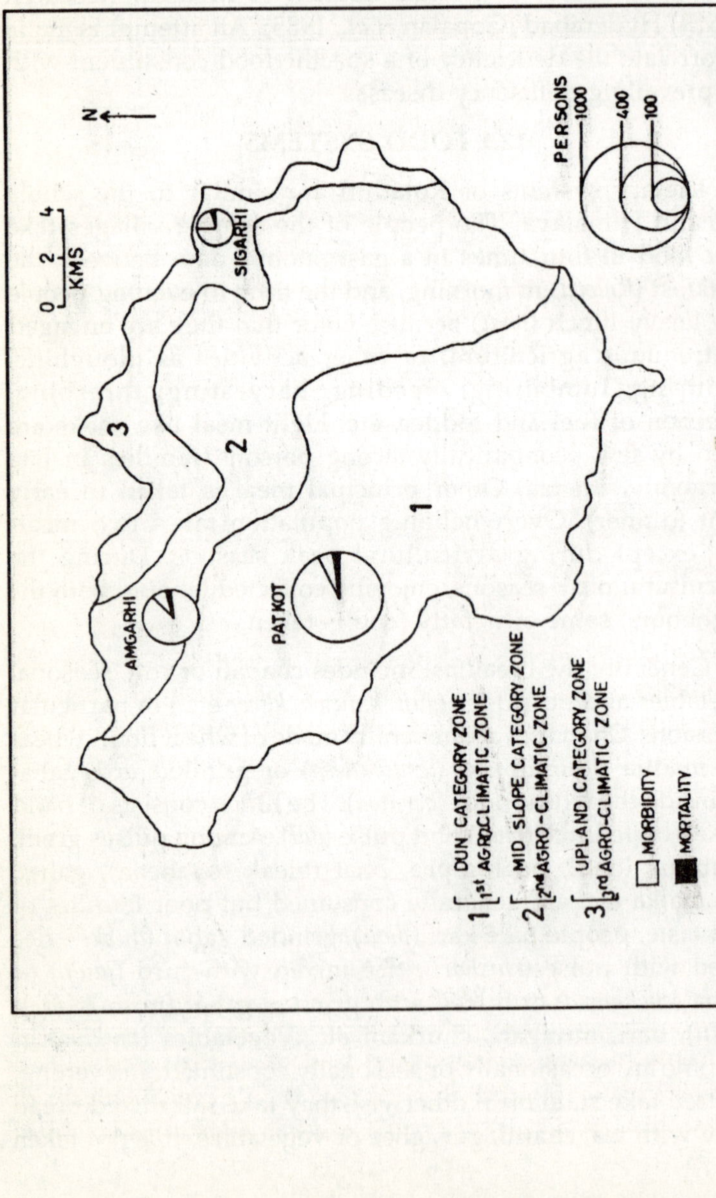

Fig. 12.1 Kotadun: Agroclimatic Zone, Morbidity and Mortality

by matured persons. People take dinner in the form of bread with vegetables. Flesh products are occasionally consumed. Curd is almost consumed with meals particularly in rainy season. Some wild vegetables or leaves are also consumed. Fried meals as puri, Kachauri, haluwa, bare, kheer etc. are taken on the occassions of festivals and other ceremonies. Although the food pattern is almost similar throughout the area but it varies from season to season. Anthropological characteristics of the inhabitants in the whole area are similar. Normal height of the people is 150 to 160 cm and weight 50 to 60 kg.

12.2 NUTRIENTS AND DEFICIENCY DISEASES

Dun Floor Village—Patkot

It is clear from table 12.1 that 54.83 per cent of total energy intake is available from cereals and pulses while 17.51 per cent is taken from milk, flesh and sugar. Only 27.66% energy is obtained from spices, vegetables (root and leafy), oil and ghee and fruits. Table 12.2 clears that fat, protein, carbohydrates, calcium, iron, carotene, riboflavin and vitamin C fall short from the standard requirement by 17.46, 12.21, 11.75, 13.81, 28.43, 6.20, 43.33 and 38.2 per cent respectively while phosphorus, thiamine and niacin are sufficiently present in the diet. Table 12.3 shows that 8.52 per cent population suffers from various deficiency diseases. Only 10.42 per cent of the total patients died because of the deficiency diseases. General weakness of the inhabitants has not been considered.

The close relationship between the actual intake of the various elements of diet and prevailing diseases among the village people is clear from table 12.2 and figure 12.1. The actual intake of vitamin A which is essential for growth and development of the body is only 900.51 μg as against the standard requirement of 960 μg. It means that vitamin A intake of the people is about 6 per cent less than that of the actual requirement. As a result 24 persons suffered from vitamin A deficiency diseases. About 4 cases of xerophathalmia (4.49%), each of night blindness (1.04%) and kerotomalacia (1.04%), 10 cases of angular stomatitis (10.04%) and 8 cases of dyspepsia (8.32%) were noted in 1986-87. Calcium and vitamin B_2 (riboflavin) the other deficient constituents, fall short by 13.81 and 43.33 per cent respectively. Therefore, 36 cases (37.5%) were suffering from various diseases—5 cases (5.21%) of dental decay, 4 cases (2.08%) of osteomalacia, 10 cases (10.42%) of rickets, 2 cases (2.08%) of haemophilia, 6 cases (6.25%) of scurvy, 2 cases (2.08%) of diarrhoea and 7 cases (7.29%) of constipation. Eight cases (infants) died of rickets or calcium deficiency diseases, one man of haemophilia, and 4 infants of scurvy and constipation. Similarly per head per day of iron is about 28.63 mg as against the requirement of 40 mg, with the result of this 12 persons (12.5%) suffered from anaemia, out of this 2 persons died. Although there is surplus consumption of niacin and fat from the standard requirement even than 3 cases (3.13%) suffered from pellagra, hypoprotonamia and glossitis, respectively. Similarly 3 cases of protein and niacin deficiency were suffering from kwashiorkor and carbohydrates deficient constituents which fall short by 11.75 per cent from standard requirement, thus 4 cases (4.17%) suffered from diabetes. It is also noted during the field work that there are several other causes of diseases among them, uneven distribution of nutrition elements among the village people, unscientific method of food preparation, indigestion of the village people and dull residential conditions are the main facts of several other diseases. Owing to the 38.2 per cent deficiency of vitamin C (ascorbic acid) 6 persons (6.25%) were suffering from pyorrhaea.

Mid Slope Village—Amgarhi

Table 12.1 shows the 61.15 per cent of the total energy intake which 37.82 per cent available from kharif and 23.33 per cent from rabi crops. About 25 per cent per head per day energy intake depends vegetables (leafy or root), spices, fruits, oil. Remaining 13.8 per cent energy is obtained from flesh, milk food and sweets. The consumption of fruit and vegetables is seasonal and thus it constitutes negligible part of their diet. Flesh products (meat, eggs and poultry) are occasionally consumed.

Table 12.1 Actual Per Head Per Day Food Intake (1986-87)

Food Stuff	Sample Villages								
	Dunfloor (Patkot)			Mid slope (Amgarhi)			Uplands (Sigarhi)		
	kg/year	gm/day	kcal/day	kg/year	gm/day	kcal/day	kg/year	gm/year	kcal/day
Rice	57.79	158.32	552.54	60.51	165.78	587.57	52.38	143.50	500.80
Wheat (Flour)	49.50	135.62	462.42	48.47	132.81	452.89	52.40	143.55	489.51
Madua (Flour)	22.78	62.41	204.70	27.93	76.53	251.03	14.33	39.27	128.81
Barley (Flour)	–	–	–	5.26	14.41	48.43	5.25	14.39	48.36
Gram	1.71	4.68	16.85	4.59	12.57	45.23	-	-	-
Corn	–	–	–	–	–	–	10.76	29.52	101.00
Lentil	3.95	10.82	37.11	1.22	3.35	11.56	–	–	–
Other pulses	6.57	18.00	53.82	9.20	25.20	75.35	18.81	51.54	154.11
Soyabean	1.08	2.96	12.78	–	–	–	–	–	–
Vegetables	37.10	101.64	381.15	34.82	95.39	357.71	17.70	48.48	24.00
Ginger	1.23	3.36	2.25	–	–	–	–	–	–
Spices	1.48	4.05	19.27	1.34	3.68	17.52	1.07	2.93	13.95
Lahi	6.75	18.49	166.44	7.46	20.45	184.05	3.11	8.52	76.91
Peas	–	–	–	–	–	–	5.22	14.30	48.62
Potato	–	–	–	–	–	–	17.45	47.81	40.38
Milk products	54.33	148.85	174.15	50.15	137.12	160.43	36.12	101.16	118.36
Flesh foods	8.65	23.69	35.98	5.26	15.24	23.01	8.13	22.28	33.64
Fruits	26.89	73.67	81.80	13.09	25.85	39.79	6.35	17.39	10.26
Sweets	19.48	53.37	212.40	13.49	36.96	147.10	12.00	32.68	130.86
Total	–	–	2413.66	–	–	2392.07	–	–	1925.61

Table 12.2 Average Per Head Per Day Food Supply (1986-89)

Elements of diet	Standard requirement	Sample Villages								
		Dunfloor (Patkot)			*Midslope (Amgarhi)*			*Uplands (Sigarhi)*		
		Actual intake	*-Deficiency +Surplus*	*Percentage Departure*	*Actual intake*	*-Deficiency +Surplus*	*Percentage Departure*	*Actual intake*	*-Deficiency +Surplus*	*Percentage Departure*
Energy kcal	2430.00	2413.66	–16.34	–0.67	2392.67	–37.33	–1.54	1925.61	–504.39	–20.76
Protien gm	66.0	57.94	–8.06	–17.46	58.12	–1.88	–3.13	63.95	–2.05	–3.11
Fat gm	50.0	41.27	–8.73	–11.76	39.22	–10.78	–21.56	24.51	–25.49	–50.98
Carbohydrates gm	430.0	379.46	–50.54	–11.75	382.40	–47.60	–11.07	364.49	–65.51	–15.23
Calcium mg	800.0	689.54	–110.46	–13.81	722.12	–77.88	–9.74	533.44	–266.56	–33.32
Phosphorus mg	1400.0	1556.18	+156.16	+11.16	1607.78	–207.75	+14.84	1654.18	+254.18	+18.16
Iron mg	40.0	28.63	–11.37	–28.43	29.43	–10.57	–26.43	27.74	–12.26	–30.65
Carotene µg	960.0	900.51	–59.49	–6.20	625.54	–334.46	–34.84	263.83	–696.17	–72.52
Thiamine mg	1.8	1.98	+0.18	+10.00	1.87	+0.07	+3.89	1.96	+0.16	+8.89
Riboflavin mg	1.5	0.85	–0.65	–43.33	0.86	–0.64	–42.67	0.78	–0.72	–48.00
Niacin mg	16.0	21.74	+5.74	+35.88	16.50	+0.50	+3.13	16.15	+0.15	+0.94
Ascorbic acid mg	50.0	30.90	–19.10	–38.20	23.71	–26.29	–52.58	32.63	–17.37	–39.74

It will be seen from the table 12.2 that in the village of Amgarhi, the diet of the people mostly lacks protein, fat, carbohydrates, calcium, iron, carotene, riboflavin and vitamin C. The deviation of 3.13 per cent of protein, 21.56 per cent of fat, 11.07 per cent of carbohydrates 9.74 per cent of calcium, 26.43 per cent of iron, 34.84 per cent of carotene, 42.67 per cent of riboflavin and 52.55 per cent of vitamin C from the normal requirement is rather alarming in the village dietary standard while phosphorus, thiamine and niacin are sufficiently present in the diet.

Table 12.3 shows that 11.35 per cent of the total population or 37 persons were suffering from various nutritional deficiency diseases and 5 cases (13.00%) of the patients died from these diseases.

A comparative study of table 12.2 and 12.4 shows the close relationship between the deficiency diseases and the constituents of the food consumed by the villagers. There is shortage of protein and niacin with the result of this deficiency one person suffered from kwashiorkor and 2 persons from hypoprotonamia. The actual intake of calcium is only 722.12 mg or 9.74 per cent below the standard requirement and 10 persons or about one third of the total patients were the victims of calcium deficiency diseases. Dental decay (3 cases) and rickets (15 cases) are the prominent calcium deficiency diseases. Osteomalacia (one case) and haemophilia (1 case) are the other calcium deficiency diseases in the village.

The consumption of carotene by the people here is 625.54 µg. i.e. 34.84 per cent below the standard requirement and as a result 8 persons suffered from this deficiency. Xerophathalmia (3 cases), angular stomatitis (2 cases) and dyspepsia (3 cases) are the prominent carotene deficiency diseases are noted in the village. Two persons of the deficiency of carbohydrate namely diabetes are observed. The supply of riboflavin is 0.86 mg or about 42.67 per cent below the standard requirement. As a result of this shortage 6

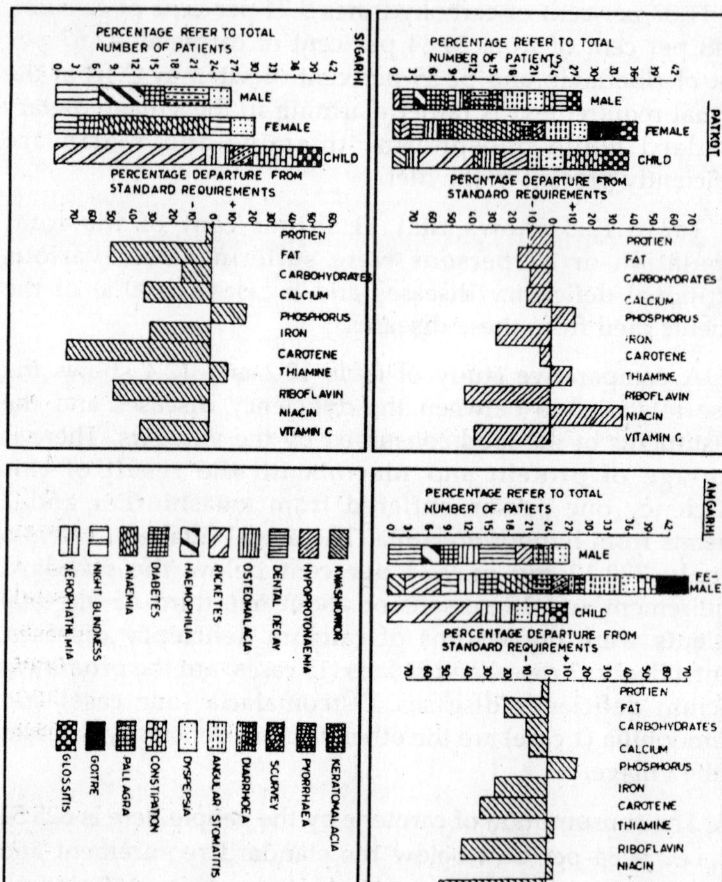

Fig. 12.2 Nutritional Status and Deficiency Diseases (1986-87)

persons suffered from vitamin B_2 deficiency diseases such as scurvy (2 cases), diarrhoea (2 cases) and constipation (2 cases). Due to the iron deficiency 2 persons suffered from anaemia disease. Three cases of pyorrhaea were also noted. Only one case suffered from glossitis due to the shortage of 21.65 per cent of fat. Two cases of goitre were also noted in the village.

Upland Village—Sigarhi

The inferior quality of soil in Sigarhi gives comparatively lower yields. Grain crops are not grown there on account of climatic extremes. The villagers fulfil their food requirements from external supplies. The per head per day consumption of energy is 1925.61 kcal which is derived from rice, corn, madua, pulses, wheat etc. (Table 12.1). The supply of other sources of food is occasional and very nominal.

Table 12.2 shows that phosphorus and thiamine are sufficiently present in the diet of the villagers. The deficiency is 3.11 per cent of protein, 50.98 per cent of fat, 15.23 per cent of carbohydrates, 33.32 per cent of calcium, 30.65 per cent of iron, 72.52 per cent of carotene, 48.00 per cent of riboflavin, 0.94 per cent of niacin and 34.74 per cent of vitamin C in the dietary standard of the villagers.

About 25.64 per cent (30 persons) suffered from various nutritional deficiency diseases, out of which 20 per cent (6 cases) of the total patients died in 1986-87 (Table 12.3). The deficiency of calcium is 33.32 per cent as a result of which 40.0 per cent (12 persons) of the total patients suffered from calcium deficiency diseases prevailing in the village. The intake of vitamin A deviates 72.52 per cent below the standard requirement. It affected 13.33 per cent (4 cases) of the total patients. Two persons also suffered from night blindness and xerophathalmia.

The consumption of riboflavin is 48.00 per cent below the standard requirement and as a result 13.37 per cent (4 cases) suffered from this deficiency, scurvy (2 cases) and

Table 12.4 Nutritional Deficiency Diseases in the Sample Villages (1986-87)
(*Percentage of Each Disease to the Total Patients*)

Sample Villages	*Total population*	*Total patients*	*Percentage of patients to the total population*	*Kwashiorkor*	*Hypoprotonamia*	*Dental decay*	*Osteomalacia*	*Rickets*	*Haemophilia*	*Diabetes*	*Anaemia*
1	2	3	4	5	6	7	8	9	10	11	12
Dunfloor (Patkot)	1127	96	8.56	3 (3.13)	2 (2.08)	5 (521)	4 (4.17)	10 (10.42)	2 (2.08)	4 (4.17)	12 (12.5)
Midslope (Amgarhi)	326	37	11.35	1 (2.70)	2 (5.41)	3 (8.11)	1 (2.70)	5 (13.48)	1 (2.70)	2 (5.41)	2 (5.41)
Upland (Sigarhi)	117	30	25.64	–	–	3 (10.10)	–	7 (23.32)	2 (6.67)	2 (6.67)	5 (16.67)
Total	1570	163	10.38	4 (2.45)	4 (2.45)	11 (6.75)	5 (3.07)	22 (13.50)	5 (3.07)	8 (4.91)	19 (11.67)

Table Contd . . .

Table Contd . . .

Night blind-ness	*Xeropha-thalmia*	*Kero-toma-lacia*	*Pyorr-hoea*	*Scurvy*	*Diarrho-ea*	*Angular stoma-titis*	*Dyspep-sia*	*Consti-pation*	*Pallegra*	*Goitre*	*Glossitis*
13	14	15	16	17	18	19	20	21	22	23	24
1 (1.04)	4 (4.17)	1 (1.04)	6 (6.25)	6 (6.25)	2 (2.08)	10 (10.42)	8 (8.32)	7 (7.29)	1 (1.04)	3 (3.13)	5 (5.21)
–	3 (8.11)	–	3 (8.11)	2 (5.41)	2 (5.41)	2 (5.41)	3 (8.11)	2 (5.41)	–	2 (5.41)	1 (2.70)
1 (3.33)	1 (3.33)	–	2 (6.67)	2 (6.67)	–	–	2 (6.67)	2 (6.67)	–	–	1 (3.33)
2 (1.23)	8 (4.91)	1 (0.61)	11 (6.75)	10 (6.13)	4 (2.45)	12 (7.36)	13 (7.97)	11 (6.75)	4 (0.61)	5 (3.07)	7 (4.29)

Note. : Figure in Parentheses refer to percentage.

constipation (2 cases) are the main diseases of this deficiency. Owing to the iron deficiency five women suffered from anaemia. Two persons suffered from pyorrhaea due to the shortage of vitamin C. Only one victim of fat deficiency disease glossitis have been noticed in the village.

12.3 MORTALITY AND MORBIDITY

The mortality and morbidity in each of the village is shown in table 12.3 and depicted in figure 12.1.

Table 12.4 Mortality and Morbidity (1986-87)

Sample Village	*Energy per head per day (kcal)*	*Percentage of morbidity to the total population*	*Percentage of morbidity to the total patients*
Dunfloor (Patkot)	24.13.66	8.52	10.42
Mid slope (Amgarhi)	2392.62	11.35	13.51
Uplands (Sigarhi)	1925.61	25.64	20.00

It is clear from table 12.4 that in the village of Patkot showing highest energy supply of food, 8.52 per cent of the total population suffers from various deficiency diseases and 10.42 per cent of the total patients expired of these diseases during 1986-87. Variations in production and consumption of foodgrains results in prevalence of different deficiency diseases from one place to another. Table 12.4 also incorporates that 11.35 per cent persons of Amgarhi suffer from various diseases, out of which 13.51 per cent victims died there from Sigarhi marks 25.64 per cent patients out of which 20.00 per cent expired of these diseases and the rates of mortality and morbidity increase, on higher places.

Conclusion

The study of nutrition and related deficiency diseases in the sample villages, reveals that the deficiency of calcium, fat, carbohydrates, protein, carotene, riboflavin, iron, vitamin

C is fairly high in the diet of all the villagers while phosphorus, thiamine and niacin are sufficiently available in their diet (Pant, 1988 and Pant *et al.*, 1989a).

Dietary habits in the sample villages are determined mainly by locally available food stuffs. The food is taken usually without knowing the nutritional importance. Generally villagers are unknown of their deficiency diseases from which they suffer and even they do not take those seriously as disease of dental decay, pyrrohea, constipation, dyspepsia even xerophathalmia are not seriously considered as diseases by most of the villagers.

The diet is inadequate in more than eight elements almost in all the selected villages. As a result 168 persons (10.38%) of the total population suffer from several deficiency diseases.

Calcium deficiency affects 43 persons or about one fourth of the total patients. Dental decay (6.75%) osteomalacia (3.07%), rickets (13.50%) and haemophilia (3.07%) are the main calcium deficiency diseases noted in these villages. Calcium is required as a building materials for strong bones and teeth. Without calcium the concentration of the heart would not be proper, the muscles would not contract properly to make the limbs move and the blood not to clot, calcium is an essential element for several life processes. It is most essential for growing children and pregnant or lactating women.

The deficiency of carotene of vitamin A exists in all three villages where 36 persons (22.08%) of total patients suffered from the vitamin A deficiency. Owing to its deficiency night blindness (1.23%), xerophathalmia (4.91%), kerotomalacia (0.61%), angular stomatitis (7.36%) and dyspepsia (7.97%) are the main diseases. In the absence of adequate intake of the vitamin, the outer lining of the eye ball loses it's usual most white appearance and becomes dry and wrinkled. Redness and inflamation of the eye and gradual loss of vision, may fallow. The central portion of the eye (cornea) may loss its transperance and becomes opique and soft and it is not treated in time may lead to total blindness. Vitamin A is also necessary to enable clear vision in dim light (Gopalan *et al.*, 1985).

Riboflavin or vitamin B_2 one of the most constituents of the diet, is concerned with several oxidation processes inside the cell (Ouigley, 1959). Some of the symptoms usually attributed to an adequate supply of this vitamin in the diet are the soreness of the tongue, cracking at the angles of the mouth, redness of the eyes, burning sensation in the eyes and scaliness of the skin in the region between the nose and the angles of the skin in the region between the nose and the angles of the lips (Gopalan *et al.*, 1985). Angular stomatitis and glossitis are the main deficiency diseases of this while scurvy (6.13%), diarrhoea (2.45%), constipation (6.75%) are the other related deficiency diseases.

A well balanced diet for growing child and adult should contain sufficient amount of nutrients to meet the iron requirement of the body and to allow for possible the regional and seasonal variation in the iron content of foods. Although the body needs very small amount of iron, about 80 per cent of it is found in the haemoglobin—the subsistance giving red colour to the red blood corpusles in the blood. In the sample villages, the iron consumption is generally low. About 19 (11.67%) cases of anaemia (a condition in which the haemoglobin content of blood is low) are observed due to the iron deficiency. It is worthnoting point here that particularly women are prone to suffer from anaemia in this region (Pant, 1988 & Pant *et al.*, 1989b).

The deficiency of ascorbic acid (Vitamin C) may lead to haemorrhages in various parts of the body, fragility of bones, swellings of joints, imperfections of the teeth. As a result of this deficiency 11 persons (6.75%) of the total patients are suffering from vitamin C deficiency diseases. Pyorrhoea is the main diseases which affects 11 persons. Scurvy is the other Vitamin C deficiency disease.

Protein also play a vital role in all the life process. The symptoms of its deficiency in man varied, but the early symptoms are loss to weight, latitudes, carry fatigability, decreased resistance to disease and prolonged convalescence (Heinj, 1959). It's deficiency also leads to diminished resistance to infection (Davis, 1969). However, the main diseases associates with the deficiency of protein, are the kwashiorkor and hypoprotonimia which together affects 8 persons (4.90%) of the total patients.

Due to the deficiency of carbohydrates in the diet of villagers, the diabetes disease occurs in five villages and affected 8 persons (4.91%) of the total patients. Carbohydrates form the main source of energy to the body. The element iodine has been the subject of considerable study because goitre a disease characterised by swelling of the thyroid glands in the neck and resulting from deficiency of iodine. Among the iodine deficiency disease goitre is the main disease and 5 cases (3.07%) are noted in the two villages. Glossitis and pellagra are also deficiency diseases due to the deficiency of multinutrients and even fat, niacin and riboflavin deficiency constituents 4.90 per cent of the total patients.

Owing to the poor economic conditions and illiteracy of nutritional importance, it is very difficult for a large number of people to afford protective food and to make up the nutritional deficiency. Either under-nutrition or malnutrition or both, along with ignorance and certain unscientific dietary customs, bad housing, poor drainage system, attachment of cattlesheds, poor hyginic conditions have led occupance of various diseases in the region.

The diet of the people is insufficient in many respect. It fails to supply the appropriate quantity of nutrients, this is ill balanced. The average composition of an improved diet is suggested in table 12.5 based on the recommendation of experts of Indian Council of Medical Research (ICMR) Hyderabad (Gopalan *et al.*, 1985).

Table 12.5 Average Composition of An Improved Diet Per Head Per Day (Gopalan *et al.*, 1985)

Food	*Amount (gm)*
Cereals	200
Millets	200
Pulses	70
Vegetables (root & leafy)	145
Fruits	57
Milk	170
Sweets	57
Vegetable oil, ghee, etc.	28
Flesh foods	28

On the basis of compandium table 12.6 the nutritional requirement may be fulfilled from the remedial sources to prevent one self form the deficiency diseases and thus it is a natural treatment (Pant *et al.*, 1989 b). The table 12.6 also incorporates the causes of diseases and their remedial or preventive measures. Although some of the nutrients may be obtained from the ready made food stuffs such as butter, ghee, milk, flesh, eggs etc. which are much expansive and beyond to purchasing capacity of the rural people even than the alternatives may be the green leafy and root vegetables fruits, millets, pulses, etc. which can be grown easily in the villages, or may be obtained from some wild vegetables. Madua is rich in calcium, supplies 34 times more than rice and 8 times than wheat (Gopalan *et al.*, 1985).

However, most of the deficiency diseases, affected persons do not kill the diseases unless they take a serious form. But they cause lasting physical and mental debilitation which undermine health and efficiency resulting in constant misery and suffering. As a result if which agricultural activities also suffer to a great extent (Pant, 1988).

Table—12.6 A Co mpandium on Nutrition and its Related Aspects (Pant *et al.*, 1989 b)

Sn.	*Nutrients*	*Standard requirement*	*Utility*	*Main deficiency disease*	*Symptoms*	*Remedy (Source)*
1.	Protein gm	66.00	It is body-building foods (essential amino-acids)	Kwashiorkor	Stunting growth, discolouration & sparness of hair, discolouration and peeling off the skin, sweeling of the body especially in the regions of legs, hands, and apathy.	Animal flesh, fish, eggs, soyabean, nuts, pulses, rice, wheat, etc.
2	Fat gm	50.00	To prevent the dryness of skin	Phynoderma (Zeroderma)	Dryness of skin, rough skin thich hormny papules of the size of a pin-head erupt in the certain part of a body such as buttocks, arms and trunk.	Hydrogeneted oil (Vanaspati) oil seeds, nuts, soyabean, pear.
3.	Carbohydrates gm	430.00	Main source of energy to body	Diabetes	Weekness, loss appitite excessive sugar in the urine.	Sugar, creals.
4.	Calcium mg	800.00	Building materials for strong bones and teeth	Dental caries, Osteomalacia, rickets, haemophilia, pyorrhea	Bones are softened bleeding of women, pus and bleeding in gums.	Milk, chees, green leafy vegetables, millets-madua and jhungora
5.	Phosphorus mg	1400.00	Essential for assimilation of carbohydrates and fat in the body	..	..	Cereals, pulses, nuts, oil seeds.

Table 12.6 Continued . . .

Table Contd.

6. Iron mg	40.00	It is useful for growing child & pragnent women	Anaemia	Bloodlessness, unhealthy paleness of the body	Green leafy and root vegetables, madua, jhungara, bajra.
7. Carotene µg (Vitamin A)	960.00	Necessary to keep the several epithitial tissues in the body	Eye problems, night blindness	Dim sight in the night the transparent covering over the white portion of the eye balls, losses of its transparency, slow growth of body.	Butter, ghee, curd, eggs, vegetables leaves, ripe mangoes, papaya, tomato and pumpkins
8. Thiamine mg	1.80	It is concerned with the proper utilisation of carbohydrates in the body otherwise sugar, starch adversely affected	Beri-Beri, constipation, diarrhoea, dyspepsia, glossitis	Water filling between the bones, stomach problems such as indigestion, constiveness looseness of bowels	Outer layers of the rice, wheat and other cereals of groundnut oil etc.
9. Riboflavin (Vitamin B_2) mg	1.50	It is concerned with the several oxidation process inside the cells	Eyes, tongue and face problems	Soreness of tongue cracking of the angles of the mouth redness of eyes, burning sensation in the eyes, scaliness on the skin of face	Milk, curd, chees, eggs, liver, green leafy vegetables.
10. Niacin mg	16.00	It is concerned with several metabolic reactions	Pollegra, diarrhoea, Hypoprotanamia	Skin diseases, leprosy, looseness of boweis, a nurvous disorder characterrized by anesthesia	Groundnuts, cereals, pulses, nuts and meat.
11. Ascorbic acid (Vitamin C) mg	50.00	It appears to be necessary for proper classification of bones & teeth, and it facilitates absorption of iron by keeping the iron in reduced form.	Scurvy pyorrhoea	Blood diseases, pus in the gums	Fresh fruits, green leafy vegetables, sprout or germinated pulses, gram, amla.

Chapter—13

Physical Factors and Landuse

It attempts to visualise the spatial variations in agricultural landuse caused by the differences of physical factors-lithology, climate, soil, absolute and relative relief and average slope. Under the identified three terrains the impact of physical factors on major landuse ... forest, area not available for cultivation, cultivable waste area and net cultivated area ... is computed on the basis of Karl Pearson's correlation technique.

The problem of feeding to the ever increasing population is to be solved by raising the productivity of the land without affecting the environment and by bringing appropriate change in the traditional agricultural pattern. The land productivity and the agricultural pattern are influenced by a number of factors, the major ones being physical, socio-economic and technological. Since the transformation of physical condition is beyond the capacity of man, one should therefore, try to control the socio-economic requirements by adjusting these with the physical environment by following technological know-how.

An attempt is made here to interpret the regional variation and impact on the existing land utilization on the basis of broad categories of slope, relative relief, absolute relief and lithological characteristics. The area is divided into three types of terrain—(i) Level area (ii) Mid slope (iii) High slope. For finding out the effective factors, correlation analysis of physical and cultural aspects is made using of Karl Pearson method. The study is made in the three steps:

(i) On the basis of physiographic characteristics all the villages are grouped into three physical strata,

(ii) The physical and the cultural environment of each of the physiographic units is shorted out, and

(iii) Correlation is computed with the help of statistical measures.

13.1 PHYSICAL STRATA

1. Level Area

It encompasses an area of 118 km^2 or 32.7 per cent of the total area. Net 72.45 per cent of this terrain is under the control of Forest Department and remaining 27.55 per cent area comprising of 35 villages, is governed by Revenue Department. Landuse categories indicate 3.75, 8.35, 16.0 and 71.9 per cent area respectively for forest, area not available for cultivation, cultivable are and cultivated area.

Ranging between 530 and 1140 m., the absolute relief is 658/km^2. Towards southern and northern parts dissimilarities of relief are observed but the middle part is almost level. The relative relief ranging between 20 and 340 m., indicates high susceptibility to erosion owing to loose lithological traits. The mean value of relative relief is 56.86 m/km^2. The Average slope ranges between 0° and 17°. It increases with the increasing order of height but covers comparatively lesser area. The average slope is 5.3^o/km^2.

2. Moderate Slope Area

It comprises an area of 140.78 km^2 or 39 per cent of the total area. Total 75.7 per cent of this terrain is under the Forest Department and remaining 24.32 per cent area is under

Revenue Department. The number of villages lying therein is 43. Major landuse categories indicate 18.4, 5.50, 56.3 and 19.8 per cent area under forest, area not available for cultivation, cultivable and cultivated area respectively.

Ranging between 860 and 1700 m., the mean value of absolute relief is 1234.9 m. The relative relief with an average value of 356.3 m. ranges between 100 and 680 m. The average slope ranges from 17° to 40° the mean value of which is 27.5°.

3. High Slope Area

It comprises an area of 102 km^2 or 28.3 per cent of the total area. Forest Department ownes 48.48 per cent area of this terrain while 51.52 per cent area is under Revenue Department or 18 villages. Major Landuse categories include the area of 5.30, 3.60, 75.20 and 15.90 per cent under forest area not available for cultivation, cultivable area and cultivated area respectively.

The absolute relief with a mean value of 1747.5 m. ranges between 1400 and 2080 m. Relative relief ranges between 300 and 580 m. with an average of 443.75 m. Average slope ranges between 23° and 40° the mean value of which is 31.28°.

13.2 PRODUCTION

The production of the paddy is decreasing with the increasing order of slope and height but that of the madua is in receprocal order. For optimization of agricultural production, one should therefore, shift the cultivation of paddy towards valleys and that of madua towards mountainous parts. According to altitude the decreasing order of production is of soyabean, pulses, wheat, gram, lentil and lahi while the increasing order is of madua, corn and barley.

Table 13.1 Kotadun: Production (in qtl./ha.)

Crops	*Valley*	*Midslope*	*High Slope*
Paddy	25.5	18.4	11.4
Madua	4.6	5.3	6.0
Jhungora	-	7.0	5.6
Corn	7.2	8.2	9.2
Soyabean	17.8	12.8	7.8
Kharif pulses	5.8	4.2	2.6
Wheat	14.2	10.7	7.2
Barley	8.8	7.6	8.4
Gram	6.0	4.2	2.4
Lentil	4.1	3.0	1.9
Lahi	5.1	3.4	1.7

13.3 CORRELATION

Population

It is most important to note that the settlements once were towards the northern boundary of the dun floor. Later on these shifted towards level area of the Dun. Probably the level area was densely forested in the past and was the home of dangerous wild animals. Now the forest cover has been converted into cultivated farms but the position of the northern part is same as it would have been in the past. Provided with almost all the modern facilities, the level area is densely populated. The total population according to the census 1981, is 10966 with a density of 337 persons/km^2.

Towards the midslope where the hazardous environmental conditions are less suitable for settlement, the density of population decreases to 110 persons/km^2. The uplands characterised by high slope and altitude, have old settlements and the population density increases to 238 persons per village.

The correlation study (Table 13.1) shows that the population is decreasing where the average slope, relative relief and dissection index is increasing thus the correlation values make their place in negative set of matrix. In mid slope areas, the average slope seems to be the controlling factor of population density but the other physical factors are inactive there. Relief characteristics have clear impact on population in high slope area. All values of high slope indicate negative relation with population. Towards uppermost part forest cover is decreasing and the population is increasing. The main cause of decreasing forest cover is the increasing human activities and their dependence on primary sector of economy. Total population and cultivable area is positively correlated. As the altitude is increasing its correlation is becoming strong. Cultivated area and population are increasing towards high altitude therefore correlation is becoming strong.

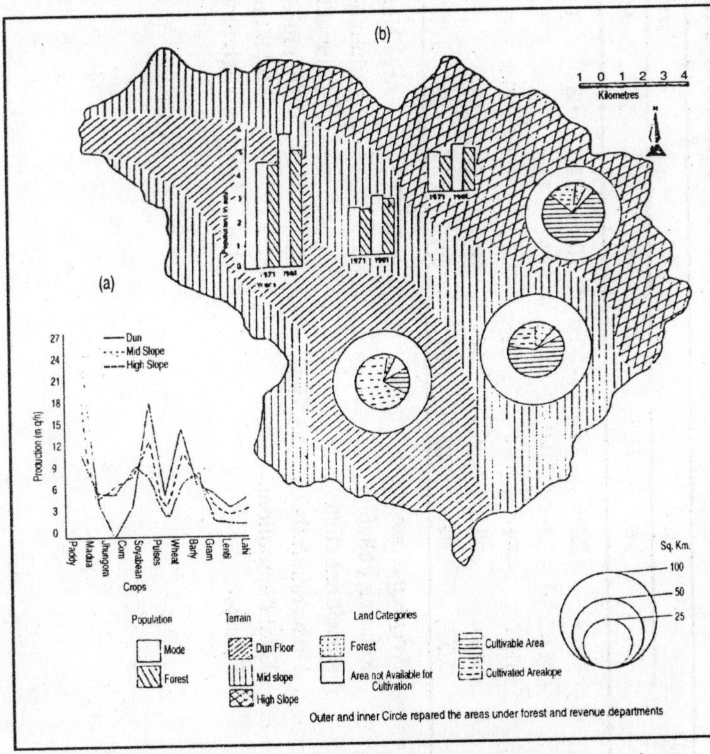

Fig. 13.1 (a) Crop Production Graph (b) Terrain: Landuse and Population

Table 13.2 Kotadun: Correlation of Physical Properties and Landuse

Variables	As.			Rr.			Ab.r.			D.I.		
	D.	M.	H.	D.	M.	H.	D.	M.	H.	D.	M.	H.
T. P.	–.16	–.24	–.34	–.20	.10	–.21	.06	.04	–.23	–.18	.04	–.34
F.	–.042	–.03	–.04	.65	.13	.18	.32	.22	.47	–.36	–.002	–.26
A. N. C.	.10	–.04	–.44	.025	–.5	.58	.34	.53	.31	–.01	–	–.11
C. A.	.46	–.05	–.46	.64	.27	.15	.65	.55	–.50	.50	.02	–.13
Ct. A.	.005	–.10	.08	.063	.09	–.27	.12	.20	–.16	–.06	.04	–.39

As. – Average Slope
Rr. – Relative Relief
Ab. r. – Absolute Relief
D. I. – Dissection Index
T. P. – Total Population
F. – Forest
A. N. C. – Area Not Available for Cultivation
Ct. A. – Cultivated Area
C. A. – Cultivable Area
D. – Dun Floor
M. – Mid Slope
H. – High Slope.

Forest

From dunfloor to upland the forest cover is in decreasing order. For ecological balance this position is not good and effective social forestry programme are required to be implement towards the upland areas with the coordination between socio-economic and environmental restoration needs.

Keeping in mind this fact three plantation models in Patkot and Saunjala are developed under the extension programme of the project. The distribution of forest area ranges between 0 and 118.43 hectares per village in Dunfloor. In mid slope area per village forest cover ranges between 0 and 214.96 ha. In high slopes the range of forest cover is between 0 and 244.30 hectare per village. As the average slope is increasing the forest cover is decreasing. Relative relief and absolute relief do not seem effective. In dunfloor and mid slopes the cultivated area and the forest cover is increasing simultaneously but in uplands it is in reciprocal order. This fact is worth to note particularly for the purpose of checking the deforestation process.

Area Not Available for Cultivation

The range of area distribution of this category is between 0.62 and 60.31 hectares per village. Because of increasing settlement activities of constructing road, canal house, institution and marketing centres a sufficient proportion of land is coming under this category use. Towards the southermost part of dunfloor the settlements are rare and the area under this category use is very nominal. In midslopes the range of this category area is between 46.65 and 1256.07 hectares per village. It in increasing order along with the higher elevation.

Cultivable Land or Waste Land

A piece of land which does not yield a return to the cultivators is known as wasteland. Although this land was used once but at present cultivators are unable to get any output with reference to input amount in it. If the tendency of measuring input and output remains continued a reasonable proportion of land may come under wasteland. Attentioning such tendency present work is formed for the optimization of landuse for environmental restoration. Keeping in view such problem reclamation is necessary. But before doing anything it is fundamental need to investigate its proper use.

The cultivable area is increasing along with the absolute relief, relative relief, average slope and dissection index. (Table 13.1). It is worthy to note that the population is increasing as the cultivable area goes on increasing. At present the lackness of fertile land and small size of land holdings are diverting people to other means of livelihood. Therefore increasing population does not seem able to decrease the wasteland.

Cultivated Area

Table 13.1 indicates that the relative relief, absolute relief and dissection index increasing from Dunfloor to highlands along with increased cultivated area. Total population is positively correlated, therefore from dunfloor to highlands where population is high, the cultivated land is also high. The impact of forest on cultivated land is positive and from dunfloor to highland it is becoming on stronger. In dunfloor and highlands, the area not available for cultivation includes the houses, roads, canals, institutions and markets therefore there is positive correlation with cultivated land. In midland it is negative because the area not available for cultivation includes active and old landslides, slumping of land, deep gorges and steep slopes which are not the supporting factors of the cultivated land.

Overall observations may be concluded by giving some prospective suggestions:

1. Level area should be used for intensive agriculture. The irregation capacity should maintained by sizeable pipe lines to escape the wastage of water.
2. Intermittent unstable and fragile zone of the study area may be used for social forestry following contour terraces to meet the twin demand of fuel and fodder.
3. Upland areas may be used for growing the cool temperate fruits and vegetables.

Chapter—14

Action Programme for Environmental Restoration

Patkot, Aonlakot, Bajuriahaldu, Amgarhi, Nalni and Sigarhi villages are selected for examining the trends of landuse prevailing in Kotadun. Besides these, two villages are taken under intensive action programme launched for alround eco-development and restoration of damaged environment, optimal, multiple and wasteland development simulated landuse models required for the environmental restoration in Kotadun, are developed in these villages.

14.1 ENVIRONMENTAL PROBLEMS AND ACTION STRATEGIES

Unstable structural traits, gritty nature of fragile rocks, heavy rains and increasing human activities are responsible for environmental hazards and land degradation by increasing rate of erosion in the environs of these villages and multiplied rate of siltation and flooding in the contiguous Bhabar. The problem calls for immediate actions and intensive core to be treated at the earliest. Since the land degradation the most serious environmental problem in the Himalaya, lies on landuse optimization and development of public wasteland, village Patkot is selected for special study and two of the sites affected by the problem, are taken under intensive care and immediate cure programme. Observing some positive responses the third site is developed in village Saunjala.

Working site-1 with an area of 3.58 ha, is the poor quality agricultural land just below the newly constructed P. W. D. road in Patkot. It is taken under cultivation during the last thirty years. The terraces are recent and are built in unplanned way. The field units are uneven and sloping. Problems of soil erosion and terrace- falls are frequent. The extension programme launched at this site are mainly multiple landuse oriented for introduction and development of intensive agriculture, horticulture and farm forestry.

Working site-2 is the degraded public waste land owned by State Revenue Department. It is at the southern end of Patkot. Trees are sparse and stunted but the growth of lantana and otherwise unwanted shrubs is luxuriant. It is the degraded forest land. About 10 ha. of this land is taken under action oriented programme for development of social forestry, cultivation of fuel, fodder and timber bearing trees, and grasses.

Working site-3 is the private waste land situated at the bank of Dabka river in Saunjala. Once cultivated, this land is degraded by erosion of Dabka river which changes its course every year in the dunfloor. An area of 1.2 ha. of this degraded land is selected for reclamation and development of hortico-farm forestry.

14.2 ACTION ORIENTED PROGRAMME

1. Education on Land Degradation and its Environmental Implication

Meetings of village folk were organised during the course of field investigations and lectures on sensitive Himalayan environment, increasing human activities in fragile zones and consequent land degradation were delivered citing local examples.

2. Education on Fast Deteriorating Condition of Community Forests and its Environmental Implications

The community forests are intended for meeting the daily requirements of fuel, fodder, agricultural implements and minor timber of the villagers. Since, no attention is paid towards their conservation and development, these forests are converted into waste lands. Since these lands are covered with lantana and other harmful shrubs, and suffer from accelerated soil erosion, the villagers now depend on contiguous reserved forests for their daily needs. Degradation of reserved forests is also a common phenomenon. Therefore, the villagers were made aware of the immediate need of conservation and development of farm and social forestry. Most of the villagers are now planting their waste lands.

3. Creation of Public Awareness for Environmental Restoration

During field work the research and the field staff created public awareness of immediate need of restoring the degraded physical environment, for adopting multiple landuse and raising fuel, fodder, fruit and timber bearing plants on field raisers and in waste lands. Introduction of hortico, farm and social forestry will make the rural society less dependent on the distant forests for meeting the daily needs of fuel, fodder, timber on the one hand and conserve the soil, time, energy and money on the other.

4. Demonstration on Intensification of Agronomic Eco-System and Implementation of Multiple Landuse

In order to execute these programme working site-1 was selected. This site was aimed at developing a simulated optimal landuse model for alround agronomic eco-system by the way of introducing scientific agricultural techniques to the owner cultivators and demonstrating these techniques on the field for implementation. The demonstration programme launched at this site, include proportional distribution of area among different category uses for agricultural land, reorganisation and levelling terraces for soil conservation, construction and maintenance of retaining walls, lining of fast flowing drains, use of improved variety seeds, chemical fertilizers and tools, implementation of line-sowing strip cropping, shower irrigation and crop rotation methods, and introduction of hortico-farm forestry.

Table 14.1 Area Under Demonstration Programme at Site-1

Plot No.	*Area (ha)*	*Landuse before the programme*	*Landuse after the programme*
1965	0.024	Waste	Plantation
1966	0.168	Fallow	Coarse-grains & Pulses
1967	0.336	Coarse grains	Food grains & Pulses
1970/1	0.068	Orchards & Pulses	Orchards & vegetables
1970/2	1.110	Food grains	Food grains & Pulses
1970/3	0.102	Fallow	Coarse-grains & pulses
1971	0.96	– do –	– do –
1972	0.104	– do –	– do –
1973	0.344	– do –	– do –
1976	0.128	– do –	– do –
1977	0.344	– do –	– do –
1994	0.024	– do –	Orchards & fodders
1995	0.120	– do –	Orchards & Foodgrains
1996	0.072	Cash crops	Cash crops
1997	0.064	Settlement	Settlement & kitchen garden
1998	0.024	– do –	– do –
1999	0.080	Food grains	Orchards & cash crops
2000	0.184	– do –	– do –
2001	0.040	– do –	– do –
2002	0.064	– do –	– do –
2003/1	0.024	Barren land	Plant nursery
2003/2	0.064	Settlement	Settlement & kitchen gardens
Total	3.584	–	–

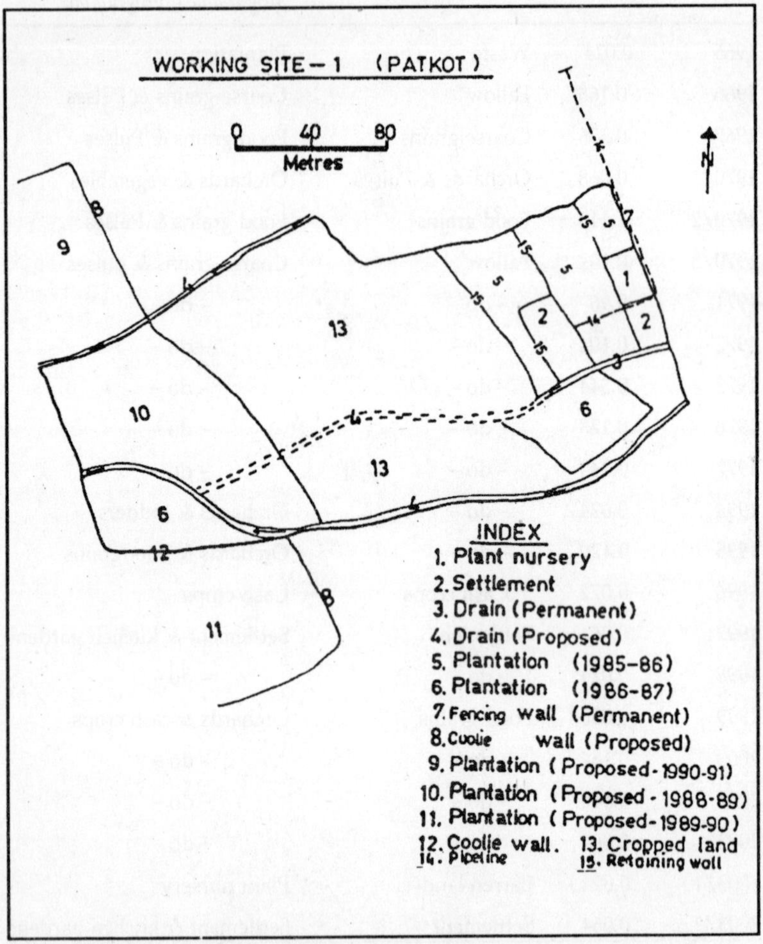

Fig. 14.1 (a) Crop Production Graph (b) Terrain: Landuse and Population

5. Demonstration on Development of Public Waste Land and Introduction of Social Forestry

These programme were extended at working site-2 and 3. Degraded public and private waste lands were restored and developed for introduction of social and farm forestry. The programme launched at these sites include clearing of approach lines, mapping of the sites, construction of fencing walls, cutting and uprooting of lantana and other harmful shrubs, digging of pits, planting of fuel, fodder, fruit and timber bearing plants, weedings and replacement of dead plants.

6. Raising of Seedlings

In order to meet the requirements of plant seedlings for the plantation programme, replacement of dead plants and supplying to the villagers to encourage tree culture, a plant nursery at the working site-1 was developed. A patch of non-agricultural barren land was reclaimed for this purpose. Seedlings of plant species indigenous or exotic, of economic importance as well as the conservational value were raised through seeds or cuttings. A lead pipe line and a water storage tank were also got developed for irrigation of the nursery site.

7. Creation of Employment

The programme launched at these three sites, created part time employment to the villagers. The cultivators during their leisures, were employed as unskilled and semi-skilled labourers in these programme.

PROGRESS OF ACTION PROGRAMME

The year-wise progress and performance of the action programme performed at three working sites of the project is given in table.

First Year (1985-86)

Two plantation models were developed in Patkot. One of these models is in private land and the other is in public land. During the year 1985-86 a number of 202 saplings were planted at site-1 and 3200 saplings at site-2. The servival rate of these saplings was 63.4 per cent and 54 per cent respectively. A plant nursery was also developed at site-1 in which a total of 460 seeds of various plant species were sown. The performance of the nursery was 300 healthy saplings.

Second Year (1986-87)

In 1986-87, 1300 saplings including those obtained from the project plant nursery, were planted at site-1, the servival rate of which was 81.6 per cent. A total of 2625 seeds of various plant species were sown in the nursery. The performance of the nursery was 1735 saplings. It being a drought year, the plantation work at working site-2 was suspended.

Third Year (1987-88)

Visualizing good performance at the project working site-1, an other hortico-farm forestry model of private land was developed at site-3, in village Saunjala where 500 saplings of various plant species were planted in this year. At the plant nursery total 3500 seeds of various plant species were sown the performance of which is about 300 saplings. A good number of 1673 saplings fully obtained from the project plant nursery, were planted not only at the working site-1 but also in other private lands in village Patkot. A good number of 2300 saplings were planted at site-2 and 500 saplings at site-3.

Environmental restoration programme of plantation, mass education, reconstruction of the fields, demonstration etc. were practised in all the samples on the basis of survival rate of the plantation work private land is found more suitable. Table 14.3 incorporates the construction work at all the sites during the tenure of the project.

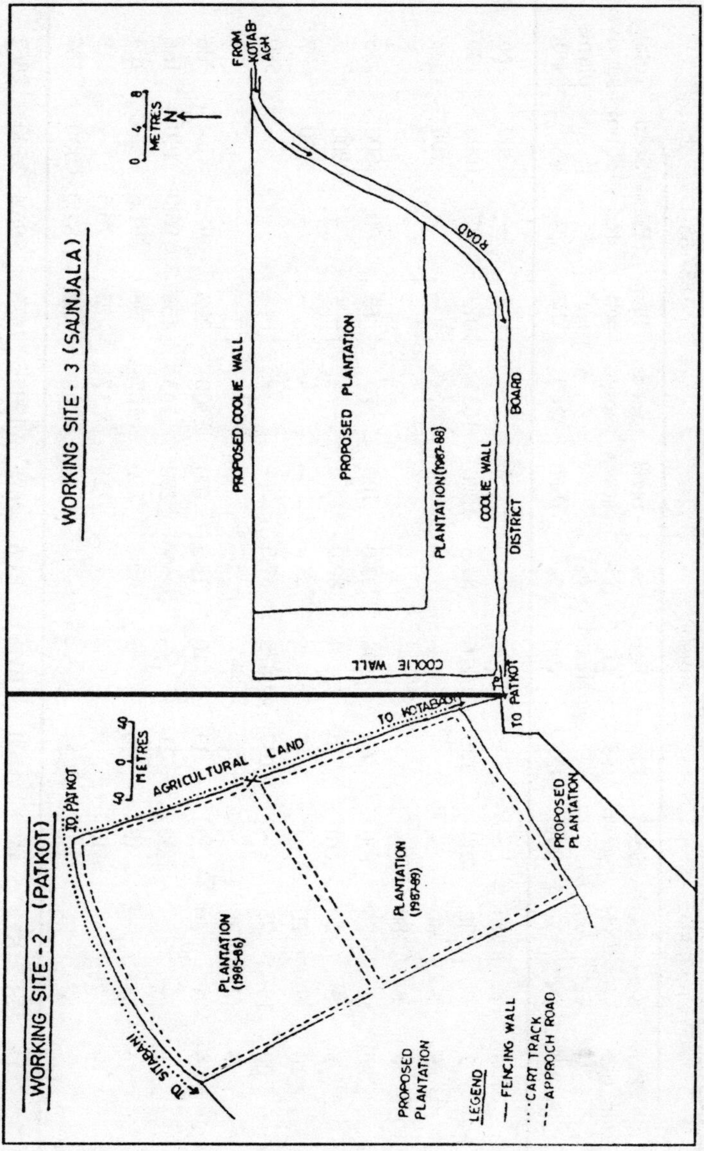

Fig. 14.2 Working Sites

Table 14.2 Plantation Work at Site—1

Particulars	1985-86				1986-87				1887-88				
	Planted (N°)	Servival (N°)	Servival %	Seed sown (N°)	Planted (N°)	Servival (N°)	Servival %	Seed sown (N°)	Seed sown (N°)	Planted (N°)	Planted %	Seed sown (N°)	Total servived plants (N°)
Aonla	100	66	66	–	–	–	–	–	–	–	–	200	66
Kathal	2	7	32	200	135	110	81.5	1000	1035	1013	97	1000	100
Nimbu	5	3	60	–	–	–	–	–	–	–	–	100	3
Peaches	10	8	80	–	–	–	–	–	–	–	–	–	8
Bhimal	20	14	70	100	60	56	93.0	100	70	63	90	1500	133
Timil	5	3	60	10	5	4	80.0	–	–	–	–	100	7
Tormil	10	7	70	–	–	–	–	–	–	–	–	100	7
Eucalyples	20	15	75	–	100	875	87.5	–	–	–	–	–	890
Silveroak	5	2	40	–	–	–	–	–	–	–	–	–	2
Teak	5	3	60	50	35	10	28.5	400	300	293	97.7	–	306
Jamun	–	–	–	30	21	–	–	200	163	156	95.7	100	156
Naspati	–	–	–	10	6	6	100.0	750	107	98	91.6	–	104
Amrud	–	–	–	10	5	–	–	25	17	13	76.5	–	13
Miscellaneous	–	–	–	50	33	–	–	50	45	37	82.0	400	37
Total	202	128	63.4	460	1300	1061	81.6	2625	1735	1673	96.4	3500	2862

Table 14.3 Construction Work (1985—88)

Sl. No.	*Particulars*	*Length (m)*	*Width (m)*	*Height/depth. (m)*	*Nature*
A.	*Site—1*				
1.	Coolie wall	23.0	0.70	1.50	Kachcha
2.	Lining of drain	53.0	1.00	.60	Pacca
3.	Fancing wall	106.0	0.45	1.50	– do –
4.	Retaining wall	21.0	0.60	1.00	– do –
5.	Pipe line	92.0	–	–	–
6.	Retaining wall	43.0	0.80	1.25	Kachcha
7.	Pits	0.45	0.45	0.60	3237 Nos.
8.	Clearing of Bushes in	–	–	–	Twice a year in 3.6 ha.
B.	*Site—2*				
1.	Coolie wall	1300.0	.75	1.25	Kachcha
2.	Line cutting	12000.0	2.00	–	–
3.	Pits	0.45	0.45	0.30	5580 Nos.
4.	Clearing of lantana bushes	–	–	–	Twice a year in 10 ha.
C.	*Site—3*				
1.	Pits	0.45	0.45	0.30	500 Nos.
2.	Coolie wall	100.0	0.75	1.25	Kachcha
3.	Clearing of lantana bushes	–	–	–	Twice a year in 1.2 ha.

Table 14.4 Plantation Work at Site-2 and 3

Particulars	1985-86			1986-87	1987-88		
	Planted (No.)	Survival (No.)	Survival %		Planted (No.) (Site-2)	Planted (No.) (Site-3)	Total plants Servived
Siras	500	100	20.0	DROUGHT YEAR 1986-87	305	45	450
Subabul	200	62	31.0		200	60	322
Bahera	100	25	25.0		125	105	225
Eucalyptis	800	500	62.5		–	–	500
Bakain	150	100	66.7		175	70	345
Aonla	50	33	66.0		259	50	342
Silver oak	500	200	40.0		141	–	341
Bhimal	100	50	50.0		335	75	460
Others	800	667	83.4		840	95	1602
Total	3200	1737	54.0		2380	500	4617

Chapter—15

Findings and Recommendation

15.1. ACHIEVEMENTS

1. MBT zone is very sensitive and fragile area accelerating land degradational problem.
2. Owing to altitudinal variations, climate, drainage, soil cover and properties also vary from one place to another.
3. Erosion intensity and land-slide frequency is increasing due to high unscientific anthropogenic activities on steep mountain slopes.
4. Dunfloor incorporates subterranean drainage pattern requiring immediate water conservational measures.
5. The area possess three major altitudinal agro-climatic zones.
6. The population of the region has a remarkable increase of 22.72 per cent in the decade of 1971—81.
7. The distribution of population is very uneven-high concentration in the lower units of dunfloor.
8. Three densities—arithmetic, physiological and agricultural—computed in the present investigation, are respectively 206, 584 and 165 person/km^2.

9. The area comprises more male population than the females. The member of female population is 970 in 1971 and 936 in 1981 against 1000 males.

10. The literacy of the area is 46.36 per cent—15.36 per cent 15.38 per cent of females and 30.98% of males in 1981 while it was 34.40 per cent in 1971.

11. Net 85.39 per cent population is engaged in agricultural activities.

12. The size of settlement largely depends upon the means of nourishing and economic conditions of the inhabitants therefore, densely populated large and compact settlements lie in dunfloor while sparsely populated small villages lie in mountainous tract of the area. Size distribution of settlement reveals that 26.56 per cent population or 6106 persons are concentrated in about 70 per cent area or 67 villages while more than two thirds or 70.44 per cent of the population resides in 30.21 per cent area or 29 villages large proportion of the population is concentrated in few villages and a small proportion of that resides in many villages. Average population size of settlement in 1981 is 215 persons. There are linear, checkerboard and irregular patterns of settlement.

13. The animal population of the area is grouped in four categories of bovine (67.41%), ovine (16.21%), equine (7.93%) and poultry (8.45%).

14. The infrastructural facilities are more in dunfloor in relation to hills. The development of services and facilities of basic physical and institutional infrastructure is far below from the adequate. More than 60 per cent of the villages fall more than 4 km away from the location of these services.

15. The land of Kotadun is divided into two major categories i.e. land owned by State Forest Department or Reserved Forest (72.17%) and land

owned by State Revenue Department or Reporting area (27.83%). The reporting area consists of six category uses viz. forest (3.53%), pasture (0.38%), not available for cultivation (5.84%), orchards (0.28%), cultivated (35.23%) and culturable waste (54.74%). The cultivated area lies maximum in dunfloor and minimum on uplands and large proportion of the waste land lies in mountainous tract and minimum in dunfloor. Mainly two crop seasons—kharif and rabi exist there and the cropping pattern varies from lower to higher elevations.

16. The agricultural efficiency or productivity index is higher in dunfloor and lower in mountainous units.
17. A new model of determining land capability is formulated in this project which classifies it into six categories.
18. Land evaluation based on land capability classification, quantifies the faulty landuse in order to investigate a feasible and applicable model of optimal landuse in accordance to the suitability and capability of land.
19. Phosphorous, thiamine and niacin are sufficiently available in the diet of the villagers while calcium, fat, carbohydrate, protein, carotene, riboflavin, iron and vitamin C are below the standard requirement. As a result 10.38 per cent people suffer from various deficiency diseases. The range of mortality is from 10 to 20 per cent showing an increasing order from dunfloor to uplands.
20. Extension work launched with full co-operation of the villagers, is more successful in private lands. The results of educating and demonstrating programme are very fruitful as the private waste land is being brought under hortico farming and social forestry programme.

15.2. RESULTS OF ACADEMIC IMPORTANCE

1. Six category landuse classes are observed in Kotadun, Kumaun Himalaya which are respectively delineated as A, B, C, D, E, and F.

   A. Area under intensive cultivation with multi-cropping system.

   B. Area under extensive cultivation with double cropping system.

   C. Area under partial cultivation with mono-cropping system including horticultural activities.

   D. Area not available for cultivation as it is under uses other than cultivation, settlement, transport and communication network, quarrying and mining, industries, commercial and recreational centres, water bodies and barren lands.

   E. Area under pastoral activities, grazing grounds, grass fields, shrub and bushes producing fuel and fodder.

   F. Area under arboural vegetation including civil, panchayati and reserved forests.

2. The land capability classes are also observed in six categories which are again portrayed as A, B, C, D, E, and F. (A_1, A_2, B_1, B_2 and C_1, C_2 in model).

   A. First class agricultural land for intensive cultivation with multi-cropping pattern—foodgrains, vegetables and cash crops.

   B. Good agricultural land for extensive cultivation with double cropping pattern.

   C. Poor agricultural land for growing one or two crops—pulses or millets in kharif season but it is first class land for the development of horticulture.

D. Area which can effectively be used for cultivation but it is recommended for non-agricultural uses—road and communication network, settlement, industries, quarrying and mining etc.

E. Area suitable for growing grasses, fuel and fodder trees.

F. Area exclusively suitable for intensive forestry.

On the basis of the correlative study of landuse and land capability a nomanclature is given to the landuse evaluation model Fig. 15.1

1. Optimal Landuse: AA, BB, CC, EE, & FF.
2. Fairly Good Landuse: AB, BC, CD, DE, EF.
3. Moderate Landuse: AC, SD, CE, & DF.
4. Fairly Faulty Landuse: AD, BE & CF.
5. Faulty Landuse: AE & BF.
6. Most Faulty Landuse: AF.

The optimal landuse condition is in 7 per cent are in upland and midslope villages while it is 15 per cent in the dunfloor. The moderate landuse is in 22 per cent area in uplands, 40 per cent in dunfloor and 54 per cent in midslopes. The faulty landuse varies from 4 to 11 per cent in there physical strata (Table 15.1). The faulty landuse can be restored by filling the gaps of land capability and landuse correlation deviation.

Table 15.1: Landuse Evaluation in Kotadun (*Area in per cent*)

Evaluation Categories	*Dunfloor (Patkot)*	*Midslope (Amgarhi)*	*Upland (Sigarhi)*
1. Optimal	15.22	7.44	7.07
2. Fairly good	32.51	23.52	21.88
3. Moderate	14.18	44.96	53.98
4. Fairly faulty	7.42	13.78	11.85
5. Faulty	4.67	10.30	5.22
6. Most faulty	–	–	–

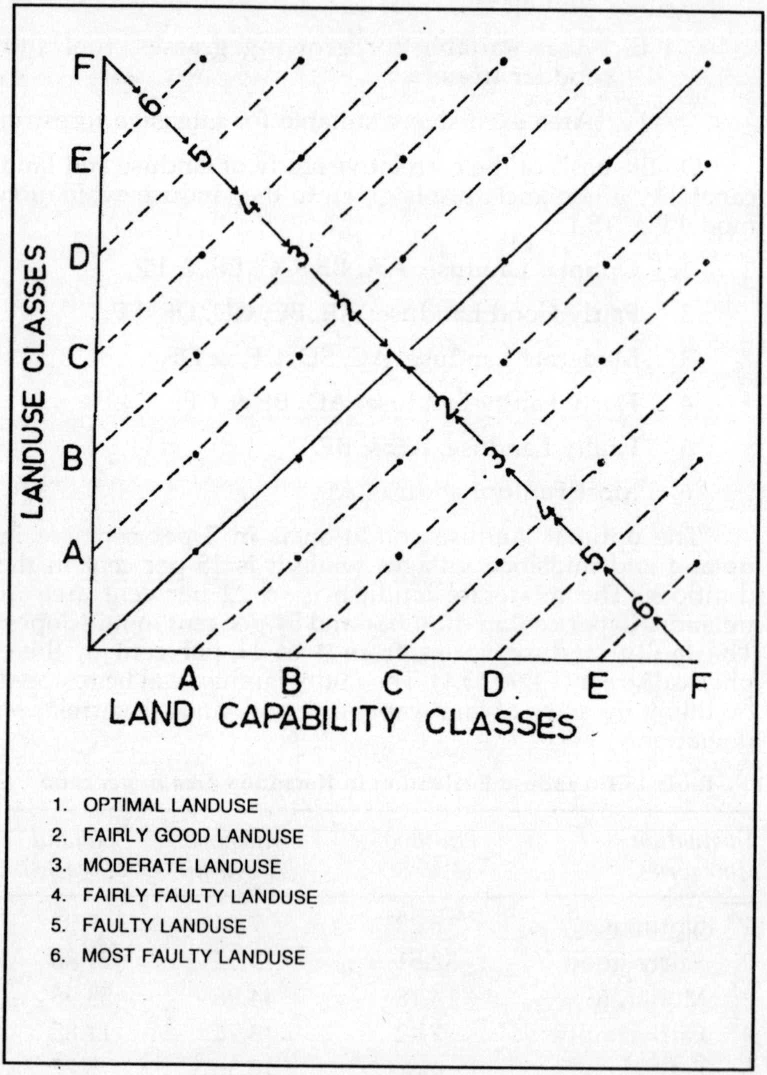

Fig. 15.1 Land Evaluation Model

15.3. RECOMMENDATIONS

The overall performance and attainment of the objectives of the research programme are thus obtained well in time during the tenure of the project. The concept of optimal landuse specially for restoration of the damaged environment as applied in this specific project, suggests two fold recommendations viz. general and specific.

General Recommendations

1. Geomorphic mapping of the project area should be undertaken in the first step for pinpointing on the environmental hazard prone zones adversely affected by soil erosion, landslide, slumping, soil creeping, earth and mud flow etc.

2. Detailed landuse survey including the changes occurred during the recent past, should be carried out in order to recording the use, misuse and overuse of each and every piece of land.

3. An appraisal of local resources comprising of water, soil, forest, mineral, population, agriculture including animal husbandry and horticulture, should be done with a view to assess their potential use without deteriorating the environment.

4. A comprehensive survey of existing infrastructural base such as transport network, means of irrigation, rural electrification, drinking water facility, educational centres, banking, and other services, post-harvests marketing processing should be carried out for formulating their prospective development schemes launched without disturbing the ecosystem.

5. Reorganization of existing faulty landuse utilizing prudently the land capability, ecological imperatives and economic and nutritive requirements of the people.

Specific Recommendations

1. The most sensitive and fragile zone of Main Boundary Fault should be restricted for anthropogenic activities such as grazing, agriculture, mining and quarrying and construction of roads, canals, buildings, dams etc. It should be preserved exclusively for growing and developing of thick vegetative cover. Activities of planting fuel, fodder and fruit bearing trees and grasses should also be intensified in this zone.

2. The water of the main streams—Bangajala, Masani, Dabka and Baur should be harnessed at the foothill zone for meeting the roaring demands of irrigational and drinking water in the dunfloor. Weirs or barrages should therefore, be constructed at the canal heads of Amgarhi, Patkot, Kota and Rurkee. Certain micro hydle stations should also be established at Patkot, Rurkee and Sitabani. The NHPC is recommended to think over launching a programme of harnessing the water of Kosi river near Betalghat, establishing an underground power-house near Amgarhi and constructing a gravity water flow tunnel from the barrage to the power house. Besides generating hydroelectricity, the water so diverted may partly be used for irrigation in the dunfloor and partly be led to the Kosi through Bangajala stream.

3. The Garjia-Sitabani-Pawalgarh forest motor road should be metalled immediately for preventing it from deep gully erosion creating havoc in the dunfloor and restraining the ecological as well as the economic development of the area as a whole. The Patalia-Kotabagh-Patkot bridle road should also be converted into a pacca light vehicular road (LVR).

4. Visualizing the performance and success of the project models developed at three sites, the fragile and sensitive mountain slopes and other affected sites of the area should be afforested intensively for growing and developing hortico-farm forestry models in the private waste lands and social forestry models in the public waste lands. For this there is an urgent need of launching intensive laboratory to land programme as submitted to the Department of Environment for necessary funding.

Bibliography

Auden, J. B., 1934: The Geology of the Krol Belt. *Rec. Geol. Surv. Ind. 67*: 357-454.

Auden, J. B., 1937: The Structure of the Himalaya in Garhwal. *Rec. Geol. Surv. Ind.* 71: 407-433.

Amani, K. Z., 1966: Land Utilization in Aligarh District. *Geographer,* Vol. V.

Ball, V., 1878: On the Origin of Kumaun Lakes, *Rec. Geol, Sur. India,* 11: 174-182.

Banerjee, B., 1955: Some Preliminary Observations on the Geomorphic features of a part of Eastern Sikkim chhangunate La-bidang Lake" *Geographical Review of India,* 17: 23-20.

Bhandari, R. K., and Gupta 1985: Problems of Landslides in the Himalaya and future directions. In J. S. Singh (Ed.), *Environmental Regeneration in Himalaya: Concepts and Strategies,* Gyanodaya Prakashan Nainital,: 39-61.

Bhardwaj, O. P., 1961: Landuse in Bist Jullundur Doab, *NGJI,* Vol. vii, Part 4.

Bhatia, S. S., 1967: A New Measure of Agricultural Efficiency in Uttar Pradesh, India, *Economic Geography,* 43: 238.

Buck, J. L., 1937: *Land Utilization in China,* London.

Clark, J. I., 1966: Morphometry From Maps. In: Dury G. H. (ed.), *Essays in Geomorphology.*

Calf, W. C., 1950: Slope Studies of Northern of Illinonis *Transcetion*, Illinois Academy of Science.

Chand, R., 1986: Settling Processes in the Himalaya. *The Indian Geographical Journal*: 63-65.

Chand, R. and M. C. Thakur, 1985: Measurement of Agricultural Efficiency in Pithoragarh District, Kumaun Himalaya. In J. S. (ed.), *Environmental Regeneration in Himalaya: Concept and Strategies*. Gyanodaya Prakashan, Nainital: 227-235.

Chattarjee, S. P., 1945: *Land Utilization in the District of 24 Parganas of Bengal, Calcutta.*

Dasmukh, V. M., 1975: Lamkede, A study in the Rural landuse. *Decc. Geog. Vol.* XIII, Part 182.

Dave, V. K. S., 1970: Major Landslides zones on Piplakoti-Badrinath and Joshimath-Malari roads, *Abstract of paper,* First seminar, Wadia Institute of Himalayan Geology, Delhi.

Davis, I. G., 1965: *Modern Public Health,* London: 57.

Day, H. C., 1961: Uttar Pradesh Forest Administration: one hundred years (1861—1961) Deptt. of Forest, U. P. Lucknow: 53.

Dayal, P., 1947: The Agricultural Geography of Bihar (unpublished Ph. D. thesis), London.

De. N. K. and Chandur, K., 1984: A Study in Soil Fertility of Balpur, The Fifth Annual Meet of I.I.G., Ahmeda-bad.

Dakuchaw, V. V., 1883: RusskiChernozem Maskva FAO/ UNDP Regional Seminar, New Delhi.

Digger *et al.*, 1974: Classification and Land-management Sym. Soil Genesis, *soil Sci.*, Calcutta.

Dove Nir, 1957: The Ratio of Relative and Absolute Altitude of Mt. Camel, *Geog. Rev.* Vol. 47: 564-569.

Enyedi, G. Y., 1964: Geographical Type of Agriculture, Budapest. *Applied Geography in Hungary.*

Fookes, P. C. and Wilson, D. D. 1966: The Geometry of Discontinuties and slope failures in Siwalik clay, *Geotechnique* 16: 305-320.

Gaur V.K. and V.K.S. Dave, 1971: Blaini tillites near Rishikesh and their Origin Jour. *Roorkee Univ.*, 13: 1-12.

Gopalon *et al.*, 1985: *Nutritive Value of Indian Foods* ICMR, Hyderabad, 204p.

Haigh, M. J., 1979: Environmental Geomorphology of Landour Mussoori Area, U. P. (Preliminary Report) *In Hima. Geol. Vol.* 9II: 657-667.

Haigh, M.J., 1984: Landslide Prediction and Highway Maintenance in Lesser Himalaya, India, *Z. Geomorphology* N. E. Suppli. Bd. S. J.: 11-37.

Heinz, H. J., 1959: *Heinz's Hand-Book of Nutrition*. New York, 33.

Hettner A., 1947: Allgenein Geographic Des Menschen *(General Geography of Humans)* Vol. I Stuttgort Hohlhammer: 6-44.

Hillman, A., 1957: *Community Organisation and Planning*, New York.

Hock, E. and J. W. Bray, 1977: *Rock Slope Engineering. Introduction of Mining and Metallurgy, London:* 402p.

Holland, T. H., 1896: Report on Geological Structure and Stability of the Hill Slopes Around Nainital. Manuscript. *Geol. Sur. India*, Calcutta.

Holland, T.H., 1908: On the Occurrence of Striated Boulders in the Blaini Formation of Simla, with a Discussion of the Geological Age of the Rocks. *Rec. Geol. Surv. Ind.* 37: 129-135.

Hukku, B. M. and P. L. Narula 1975: A Geotechnical Assessment of the Stability of Slopes at Khuni Nala and landslide at Nashri on National Highway, N. W. Himalaya Landslide and Toe Erosion Problems with special reference to Himalayan Region, *Seminar Procesling Calcutta.*: 165-174.

Jalal, D. S., 1970: Land Utilization in Bamangaon, *NGJI*, Vol. XVI.

Jalal, D.S., 1971: Land Utilization in the District of Pithoragarh (unpublished Ph. D. thesis), Agra.

Jalal, D.S., 1976: *Land Utilization in Pithoragarh*. Chaitanya ab. Allahabad.

Jalal, D.S., 1985: Environmental Degradation in Kotadun, In J. S. Singh (ed.), *Environmental Regeneration in Himalaya: Concept and Strategies*, Gyanodaya Prakashan, Nainital: 421-426.

Joshi, R. C., 1988: Quantitative Geomorphic Analysis of Kotadun, Kumaun Himalaya Unpublished Ph.D. Thesis, Kumaun University, Nainital: 188p.

Joshi, R. C., B. R. Pant, and D. S. Jalal, 1989a: A n Assessment and Distribution of Slope in Kotadun, Kumaun Himalaya. *The Deccan Geographer* 27 (1) 451—459.

Joshi, R. C., B. R. Pant, and D. S. Jalal, 1989 b: Deteriorated Aquatic Environment in Kotadun Kumaun Himalaya. In R. D. Khulbe (ed) *Perspective in Aquatic Biology*. Papsus Publishers, New Delhi: 497—502.

Joshi, R. C., B. R. Pant and A. Pande, 1990: Geomorphological Implications in Land Management. A case Study of Kotadun, Central Himalaya, *NEHU Journal of Social Sciences and Humanities 8(3)*: 15—20.

Joshi, R.C., B.R.Pant and D.S.Jalal, 1992: Landslides Along Main Boundary Fault in Kotadun : Nature, Causes and Preventions Kumaun Himalaya. In : R.K. Rai *et. al.* (ed.), *Envronmental Management* : Socio-Economic FacitsVo.1.2 Rawat Jaipur: 89-103.

Jordan, T. G., 1996: On the Nature of Settlement Geo-graphy. *The Professional Geographer, Vol.* XVIII, No. 1.

Karan, P.P., 1960: Landuse Reconnaissance in Nepal by Agro-field Techniques and Photography. *Proceedings of the American physiological Society*. Vol. 104., No. 2.

Kayaslh, S.L., 1964: Himalayan Beas Basin, *Nat. Geog. Society, India,* Varanasi: 193.

Kendal, M.G., 1939: The Geographical Distribution of Crop Productivity in England, *Journal of the Royal Statistical Society,* 162: 24—28.

Khan, A.H., 1965: Geographical Analysis of the Present Crises of Food Grain shortage in India. A paper read in school in the Department of Geograph, Nainital (unpublished).

Khanka, S.S., 1987: Role of Road Transport in Integrated Regional Development, A Case Study of Himalayan Hill Region of Kumaun in U.P. In P.S. Satsangi and A.L. Agrawal (eds), *Proceeding of Inter-national Conference on Transportation System studies,* Indian Institute of Technology, New Delhi, December 16—22, 1986.

Khanka, S.S., 1989: Road Construction and Hill Environment (with special reference to U. P.), *Journal of Rural Development, Vol.* 7 (i): 21—25 NIRD, Hyderabad.

Klages, K. H. W., 1958: *Ecological Crop Geography, New York.* The Mac Millan Co.: 44 & 335.

Kendrew, W. G., 1937: *Climate of the continent,* Oxford University Press London.

Lahiri, R., 1950: Land Utilization in Some Villages Near Jasidih. *Geog. Rev. Ind.* Vol. VII, No. 1.

Mc Harg, I. L., 1966: Ecological Determinism in Darling F. F. and Milton, J. (eds.), *Future Environment of North America, New York.* Natural History Press: 526-538.

Mc Harg, I. C., 1969: *Design with nature New York,* Natural History Press: 190p.

Midilicott, H. B., 1864: On the Geological Structure and Relationship of the Southern Portion of the Himalayan Ranges Between the river Ganges and Ravi. Men. *Geol. Surv. Ind.* 3(2): 1-212.

Middlemiss, C. S., 1890: Geological Sketch of Nainital, with some remarks on the natural conditions governing Mountain Slopes, Rec. *Geol. Sur. India,* 23,: 213-234.

Miller, O. M. and C. H. Summersoo, 1960: Slope Zone Maps, *Geographical review, Vol.* 50,: 194-202.

Nigam, M. N., 1968: Soil Erosion in Kashmir Himalaya Abstract of paper, *Int. Geol. Cong. 21st session, India.*

Oldham, B. P., 1893: Note on the Nainital landslide, *Rec. Geol. Sur. India* 13: 277-281.

Ouigley, D. T., 1969: *The National Malnutrition,* Milmarkee, 11p.

Pofali, R. M. *et al.*, 1979: *Geog. Rev. of India,* Vol. 41.

Pant B. R., 1987 a: Land Capability Classification, In D. S. Jalal (Principal Investigator) *Optimal Landuse for Environmental Restoration in Kotadun, Kumaun Himalaya* Techni. Report II, Sub. to the Deptt. of Env. (memio) and Forest Govt. of India, New Delhi (memeo): 128-37.

Pant, B. R., 1987 b: Sample Studies—Land Utilization in Patkot, Amgarhi and Sigarhi, In D. S. Jalal (Principal Investigator) *'Optimal Landuse for Environmental Restoration in Kotadun, Kumaun Himalaya'* Technical Report II, submitted to the Deptt. of Environment and Forest, Govt. of India,: 38-69 (mimeo).

Pant B.R., 1988: Land Utilization in Kotadun Kumaun Himalaya, Ph.D. Thesis Kumaun University, Nainital. p. 270.

Pant, B. R., 1992 a: An Appraisal of Land and Other Resources in Himalaya—A case study. *Indian Journal of Landscape System and Ecological Studies* 15 (2) 13-21.

Pant, B. R., 1992 b: Nutritional Status and Deficiency Diseases of Kotadun, U.P. Himalaya. A case study of Nutritional Geography. *Ind. Jour. of Landscape Systems and Ecological Studies* 15 (2) 75-82

Pant, B. R., 1994. *Food and Nutrition: A Study of Himalayan Region.* Anmol Publication, New Delhi 195 p.

Pant, B.R., 1995. Nutrition Availability Pattern of Scheduled Tribe Population in Uttarakhand-Buksha. A case, *Geog Review of India* 57 (4) 358-366.

Pant, B. R., 1996 a: Food and Nutrition Intake Scenario of Tharu Tribes in Uttarakhand Himalaya: *Indian Journal of Landscape Systems and Ecological Studies* (19): 112-119.

Pant, B. R., 1996 b.: A Geographical Study of Scheduled Castes Population in Uttarakhand (India). *The Indonesian Jour. of Geography* 28 (7) 25-38.

Pant, B. R., 1996 c: Physical Geography and Demo-graphic Structure of Uttarakhand. In K. S. Valdiya (ed.), *Uttarakhand Today*—Shri Almora Book Depot. Almora: 45-65.

Pant, B. R., 1998: *Geo-Economic Environment and Nutrition*, Anmol Publication, New Delhi. 173p.

Pant, B. R. and D. S. Jalal, 1991 a: State of Nutrition in Kotadun, Lesser, Himalaya, *Annals of the NAGI* 11(2) 64-67.

Pant B. R. and D. S. Jalal. 1991 b: Nutrition Deficiency Diseases in Indian Central Himalaya. *The Geographer*, 38 (1) 30—41.

Pant B. R. and D. S. Jalal 1991 c: Land and Infrastructural Development in Kotadun, Kumaun Himalaya. in. H. C. Upadhyay (ed.), *Modernization and Rural Development*. Anmol Publ. New Delhi 273—293.

Pant, B. R. and D. S. Jalal, 1992 a: Optional Landuse for Environmental Restoration in Kotadun Kumaun Himalaya, *CHEA Bulletins*, 5:42—51.

Pant B. R. and D. S. Jalal, 1992 b: Land Resource Management 'Model' for Kotadun, Kumaun imalaya. In J.L. Raina (ed). *Dry Land Farming in India*, Pointer, Jaipur: 86—110.

Pant, B.R. and D.S. Jalal 1993 a: Food and Nutritional Environment in Kotadun, Kumaun Himalaya. In B.P. Chaurasia (ed.), *Environmental Degradation and Global Awakening*, Chugh Publication, Allahabad: 202—222.

Pant, B. R. and D. S. Jalal, 1993. b: Altitudinal-Agro climatic Zones and Sampling Framework in Central Himalaya—A case study. In N. L. Gupta and R. K. Gurjar (eds.), *Sustainable Development* vol. II, Rawat Publication Jaipur: 159—174.

Pant, B. R. and D. S. Jalal, 1996: Environmental Hazards in Central Himalaya—A case study. In R. B. Singh (ed.), *Research in Geography: Disasters and Environment: Monitoring and Forecasting,* vol II. APH Publication, New Delhi— 291—298.

Pant, B. R. and D. S. Jalal, 1997: Recovery of Damage Environment. An Experimental Study of Indian Central Himalaya. In P. Mag, V. K. Kumra and J. Singh (eds.), *Geography and Environment* vol. I. concept—New Delhi: 162—202.

Pant, B. R. and D. S. Jalal, 1998: Changing Landuse Pattern in Indian Central Himalaya in: O. P. Singh and R. K. Pande (eds). *Human Habitat in the Mountains,* Shree Almora Book Depot, Almora.

Pant, B. R., R. C. Joshi and D. S. Jalal, 1988 a: Land Capability classification 'Model' for Kotadun Kumaun Himalaya. *Geog Ravi. of India*, 50 (1) 47—52.

Pant, B. R., R. C. Joshi and D. S. Jalal, 1988 b: Erosion in Hill Environment. *The Himalaya: Man and Nature:* 12 (7) December 18-19.

Pant, B. R., R. C. Joshi and D. S. Jalal, 1990a: Cropping Pattern and its Impact on Nutrition and Health in Kumaun Himalaya with special reference to Kotadun in M. P. Joshi, A. C. Fanger and C. W. Brown (eds.), *Himalaya: Past and Present,* Shree Almora Book Depot, Almora: 365—385.

Pant, B. R., R. C. Joshi and D. S. Jalal 1990 b: Nutritional Problem in the Kumaun Hill. An Illustration for Kotadun. *Himalaya: Man and Nature* 13 (12) 4—6.

Pant, B. R., R. C. Joshiand D. S. Jalal, 1991: Agricultural Landuse and Nutrition in Kotadun, Kumaun Himalayas. *Geographical Review of India*. 53 (4) 8—18.

Pant, B. R., R.C. Joshi and D. S. Jalal 1992 a: Agricultural Panoroma of Uttara-khand. In N. Mohammad (ed.), *New Dimention in Agricultural Geography*, Vol. 5: 81—94.

Pant, B. R., R. C. Joshi and D. S. Jalal. 1992 b: Landuse and Nutrition: The Kumaun Himalaya: in J. L. Raina (ed.) *Himalayan Environment, Man and The Economic Activities*, Vol II, Pointer Publishes, Jaipur: 287—307.

Pant, B. R. and M. C. Pant (eds.): *Glimpses of Central Himalaya. A Socio-Economic and Ecological Perspective*, Radha Publication, New Delhi: Vol. I & II, 624 p.

Pathak, H. G., 1977: A Study in Methodology and Crop Association Regions and Their Role in Agricultural Regionalization: A Case Study in U. P. Himalaya, *National Geographical Journal of India* Vol. XXIII No. 1 & 2.: 73—85.

Raisz. E. and J. Henry, 1937: An Average Slope Map of Southern New England, *Geog. Review*, Vol. 27:467-472.

Rich, J. L., 1966: A Geographical Method of Determining the Average Inclination of a Land Surface from a Contour map. *Transaction. Illinois Academy of Science*: 195-199.

Robert, J., 1920: Quoted By O. H. K. Spate and A. T. A. Learmonth (1967), *India and Pakistan*, Suffolk: 121-122.

Rupke, J.S. 1974: Stratigraphic and Structural Evolution of the Kumaun Lesser Himalaya. *Sedimentary Geology*. Vol. II: 81-265.

Rao, V. L. S. P., 1947: Social Survey and Landuse Analysis, *Ind. Geog. Jour.*, Vol. XXII, No. 3.

Roy, B. K., 1969: *Arable Uses Need Adjustment for Landuse Planning: A Case Study of the Ganga-Ghaghara Doad East*, New Delhi.

Samant, S., 1986: Landuse Planning in Mountainous Tract in Nainital, Ph. D. thesis, Kumaun University, Nainital 299.

Sengupta, P. and G. Sadasyak 1961: Economic Regionalisation of India; Problem and Approach, Monograph series, Vol I, No. 8. *Census of India*, 1961: 189-209.

Shafi, M., 1960: *Land Utilization in Eastern Uttar Pradesh*, Aligarh.

Shafi, M., 1972: Measurement of Agricultural Productivity of the Great Plain. *The Geographer* 21: 1-8.

Sing, J. and S. S. Dhillon 1984: *Agricultural Geography*, Tata McgrawHill, Delhi, 412p.

Singh, J. S., Uma Pandey and A. K. Tiwari 1984: Man and Forest: A Central Himalayan's Case study, *Ambio*: 13 (2): 80-87.

Singh, S. and R. Srivastava, 1977: A Statistical analysis of the Average Slopes of the Belan Basin. *The Deccan Geographer*, Vol. XV, Secunderabad. 297-307.

Singh, K., 1976: A Strategy for Landuse planning and Management in the Naraura Watershed in Bhikiasain Block, Almora District, paper presented at the seminar on regional/ Area planning and resource development held at Pant Nagar, District Nainital, U.P. May 27-29. 1.p

Stamp, L. D., 1948: *The Land of Britain and How it is Used*. London.

Stamp, L. D., 1960: *Our Developing World*. London.

Stamp, L. D., 1962: *The Land of Britain: Its Use and Misuse*. London.

Stamp, L. D., 1964: *Applied Geography*, Bungay.

Stamp, L. D., 1958: Measurement of Land Resources, *The Geographical Review*, Vol. XIVIII, No. 1, New York 1-10.

Stamp, L. D., 1964: *Geography of Life and Death*. London: 97p.

Strahler, A. N., 1952: Hypsometric (area-altitude) Analysis of Erosional Topography, *Bull. Geol, Soc. Am. Vol.* 63.

Strahler, A. N., 1956: Quantitative Slope analysis, *Bull. Geol. Soc. Am. Vol.* 67,: 571-596.

Smith, G. H., 1938: An Analysis of Slope, *Ann. Assoc. Ameri. Geog.*

Symons, L., 1967: *Agricultural Geography,* London, G. Bull. 32 and 95.

Tewari, A. K., 1966: Land Utilization in Jaunsar Bawar, *Dec. Geog* Vol. III & IV.

Thomas, A., 1967: *Farming in hot countries London,* Feber and Feber, 15.

Valdiya, K. S., 1967: Occurrence of Magnesite Deposits and Time Controlled Variation of Stromatolites in the Shali series, District Mahasu, H. P. Bull. *Geol. Soc. Ind.,* 4,: 125-128.

Valdiya, K. S., 1973 b: The Blaini Conglomerates of Himachal Pradesh and Garhwal Himalaya. *In Recent Researches in Geology.* Vol. I, Hindustan Pub. Corp. Delhi: 66-81.

Valdiya, K. S., 1975: Lithology and Age of Tal formation in Garhwal, and Implication on Stratigraphic Schemes of Krol Belt in Kumaun Himalaya. *Jour. Geol. Soc.* Ind., 16: 119-134.

Valdiya, K. S., 1980: *Geology of Kumaun Lesser Himalaya.* WIHG, Dehradun 291. p.

Valdiya, K. S., 1985: Accelerated Erosion and Landslide Prone Zone in the Central Himalayan Region. In J. S. Singh (ed.), *Environmental Regeneration in Himalaya Concept & Strategies,* Gyanodaya Prakashan, Nainital: 12-38.

Valdiya, K. S., 1980: Stratigraphic Scheme of the Sedimentary Units of the Kumaun Lesser Himalaya, in K. S. Valdiya and S. B. Bhatia (Eds.), *Stratigraphy and correlations of Lesser Himalayan Formations,* Hindustan Pub. Corp. Delhi. 7-48.

Vink, A. P. A., 1983: *Landscape Ecology and Landuse* (Translation of Dutch language: Landscape ecology in landgebruik): p. 264.

Vishar, S. S. 1932: Recent Trend in Geography, *Scientific Monthly Vol*. 35, p. 439.

Wentworth, C. K., 1930: A Simplified Method of Determining the Average Slope of Land Surfaces, *Am. J. Sci. Series* 5, Vol: 20.

Whitback, R. H., 1932: *The Geographic Factor New York*, Century. Co. New York: p. 87.

Worcester, P. G., 1948: *A Text Book of Geomorphology*, D. Van. Nostrand Company, New York.

Index